Cambridge Tracts in Mathematics
and Mathematical Physics

GENERAL EDITORS:
P. HALL, F.R.S., AND F. SMITHIES, PH.D.

No. 29

THE FOUNDATIONS OF
DIFFERENTIAL GEOMETRY

THE FOUNDATIONS OF
DIFFERENTIAL GEOMETRY

BY

OSWALD VEBLEN

AND

J. H. C. WHITEHEAD

CAMBRIDGE

AT THE UNIVERSITY PRESS

1960

CAMBRIDGE UNIVERSITY PRESS
Cambridge, New York, Melbourne, Madrid, Cape Town, Singapore, São Paulo, Delhi

Cambridge University Press
The Edinburgh Building, Cambridge CB2 8RU, UK

Published in the United States of America by Cambridge University Press, New York

www.cambridge.org
Information on this title: www.cambridge.org/9780521066747

First published 1932
Reprinted 1953, 1960
Re-issued in this digitally printed version 2008

A catalogue record for this publication is available from the British Library

ISBN 978-0-521-06674-7 paperback

PREFACE

THIS is intended as a companion to the Cambridge Tract No. 24, on Invariants of Quadratic Differential Forms. As its name implies it contains a set of axioms for differential geometry and develops their consequences up to a point where a more advanced book might reasonably begin. Formulae appear only incidentally and the reader is supposed to obtain those needed from the tract No. 24, or from other books and articles on the formal side of the subject.

Analytical operations with coordinate systems are continually used in differential geometry, a typical process being to "choose a coordinate system such that....'' It is therefore natural to state the axioms in terms of an undefined class of "allowable" coordinate systems, and to deduce the properties of the space from the nature of the transformations of coordinates permitted by the axioms.

The axioms for differential geometry in general are preceded by more special sets of axioms in which the structure of a space is defined by an appropriate class of "preferred" coordinate systems. Thus Euclidean geometry is characterized by the class of rectangular cartesian coordinate systems. The "preferred" coordinate systems constitute a sub-class of the "allowable" coordinate systems for any one of these spaces. The former class is small, so as to characterize the structure of the space, and the latter is large, so as to permit freedom of analytic operation.

These earlier axioms are found to be adequate for the differential geometry of an open simply connected space, the most elementary theorems of which occupy the greater part of Chaps. III–V. The more general axioms, in terms of allowable coordinate systems and without restrictions on the connectivity of the space, are given in Chap. VI. We believe that they provide an adequate foundation for any of the differential geometries which are now being studied. The complete theory which should be constructed out of these axioms would be a combination of infinitesimal geometry and analysis situs. In the final chapter we outline some of the questions which arise, in the hope that some of the readers of this tract may participate in the construction of a branch of mathematics which we are convinced is of great importance.

O. V.

J. H. C. W.

PRINCETON, N.J.

CONTENTS

CONTENTS

CHAPTER I

THE ARITHMETIC SPACE OF n DIMENSIONS

1. Arithmetic points.

Analysis habitually borrows not only terminology but also methods and results from geometry. In the present chapter we mean to indicate how this can be done without running into a vicious circle in the application of analysis to geometry. We shall be particularly concerned with the ideas clustering about the notion of linear dependence.

We shall presuppose the contents* of $Q.F.$ Chap. I, and in particular Cramer's rule ($Q.F.$ p. 6), by which a set of linear equations

(1·1) $$y^i = \sum_j a^i_j x^j$$

can be solved to yield

(1·2) $$x^i = \sum \alpha^i_j y^j,$$

provided the determinant

$$a = |a^i_j|$$

is not zero. With the notation of $Q.F.$,

(1·3) $$\alpha^i_j = \frac{1}{a} A^i_j,$$

where A^i_j is the co-factor of the element a^j_i in the matrix $||a^i_j||$, and α^i_j is called the *normalized co-factor* of a^j_i.

An ordered set of n real† numbers $(x^1, \dots, x^n)$ will be called an *arithmetic point*, and $x^1, \dots, x^n$ its *components*. The set of all arithmetic points, for a given value of n, will be called the *arithmetic space of n dimensions*. As in $Q.F.$ we shall denote an arithmetic point by a single letter x, and in considering several arithmetic points shall distinguish them by subscripts; thus x_a will stand for $(x^1_a, \dots, x^n_a)$.

In Euclidean geometry, for instance, all points are alike, but in the arithmetic space each point has an individuality of its own. In particular the points $(0, \dots, 0)$, and $(1, 0, \dots, 0)$, $\dots$, $(0, \dots, 0, 1)$ will be called

---

* This book is intended to run parallel to the *Cambridge Tract*, No. 24, by O. Veblen, called *Invariants of Quadratic Differential Forms*, which will be referred to as $Q.F.$

† There is nothing to prevent our taking numbers from any field, but we shall be content to use the real number system of analysis.

the origin and unit points respectively. We shall denote the origin by e_0 and the unit points by $e_1, \ldots, e_n$.

In many books on analysis an ordered set of numbers $(x^1, \ldots, x^n)$ is called simply a point (without any adjective). On the other hand in books on algebra* it is often called a vector. Two fundamental operations in vector algebra are multiplication by a number and addition. More precisely, if x is an arithmetic point (vector), and if a is a number, then ax is the point

$$ax = (ax^1, \ldots, ax^n),$$

while if x and y are two points, $x + y$ is the point

$$(x^1 + y^1, \ldots, x^n + y^n).$$

Combining these two operations we can define the difference of two points, and in general any linear combination,

$$t^1 x_1 + \ldots + t^k x_k,$$

of k points, $x_1, \ldots, x_k$. The theory of linear dependence has to do with properties which can be stated in terms of these two operations.

2. Linear dependence.

The points given by †

(2·1) $$\qquad x^i = t^a x_a^i, \qquad (\alpha = 1, \ldots, m)$$

are said to be *linearly dependent* on $x_1, \ldots, x_m$. According to this definition the origin is linearly dependent on any set of points. A set of two or more points is said to be *linearly independent* if no one of them is dependent on the rest. To complete the definition we say that a single point is linearly independent if it is not the origin.

A set of points, $x_1, \ldots, x_m$, is independent if, and only if, the relation

(2·2) $$\qquad s^a x_a^i = 0$$

implies $s^1 = \ldots = s^m = 0$. For if one of these coefficients s^1, say, did not vanish, (2·2) would give

$$x_1^j = -\frac{s^2}{s^1} x_2^j - \ldots - \frac{s^m}{s^1} x_m^j.$$

A relation of the form (2·1), on the other hand, is a special case of (2·2).

Let $a_1, \ldots, a_m$ be any set of points. Either $a_1, \ldots, a_m$ all coincide with

---

* E. Study, *Einleitung in die Theorie der Invarianten linearer Transformation auf Grund der Vektorenrechnung*, Braunschweig, 1923; also H. Weyl, *Gruppentheorie und Quantenmechanik*, Leipzig, 1931.

† As in *Q. F.* a repeated index will imply summation. Roman indices will invariably run from 1 to n, while the range of Greek indices will be indicated in the text.

the origin or one of them, a_1, say, is independent. Either they will all depend upon a_1 or the set a_1, a_2, say, will be independent. Proceeding in this way we shall arrive at an independent set a_1, ..., a_p, say, upon which all the points a_1, ..., a_m will be linearly dependent. We need a criterion to determine p, and we get this by considering the matrix ($Q.F.$ p. 4),

$$\| a_\beta^i \| = \begin{Vmatrix} a_1^1, & a_2^1, & \dots & a_m^1 \\ a_1^2, & a_2^2, & \dots & a_m^2 \\ \vdots & \vdots & & \vdots \\ a_1^n, & a_2^n, & \dots & a_m^n \end{Vmatrix},$$

whose columns are the m arithmetic points.

If all the $(p+1)$-row determinants ($Q.F.$ p. 9) vanish, but at least one p-row determinant does not, the matrix is said to be of *rank p*. The fundamental theorem of linear dependence is:

If p is the rank of the matrix $\| a_\beta^i \|$, the points a_1, ..., a_m are all dependent upon p of them which are themselves independent.

To prove this, first consider the case where $m \leqq n$. The matrix $\| a_\beta^i \|$ is of rank p, and without loss of generality we may suppose the determinant

$$a = | a_\mu^\lambda |, \qquad (\lambda, \mu = 1, \dots, p)$$

to differ from zero. If $p = m$ it follows from Cramer's rule for solving linear equations that the points a_1, ..., a_m are independent. For, since $a \neq 0$, the equations

$$a_\beta^\alpha x^\beta = 0, \qquad (\alpha, \beta = 1, \dots, m)$$

have the unique set of solutions $(0, \dots, 0)$.

If $p < m$ the points a_1, ..., a_p are shown to be independent by the argument we have just used. Let A_σ^1, ..., A_σ^p be the co-factors of a_1^i, ..., a_p^i in the matrix

$$\begin{Vmatrix} a_1^1, & \dots & a_p^1, & a_\sigma^1 \\ \vdots & & \vdots & \vdots \\ a_1^p, & \dots & a_p^p, & a_\sigma^p \\ a_1^i, & \dots & a_p^i, & a_\sigma^i \end{Vmatrix}.$$

The determinant of this matrix is

$$(-1)^p a a_\sigma^i + A_\sigma^\lambda a_\lambda^i, \qquad (\lambda = 1, \dots, p).$$

For $i \leqq p$ this vanishes since two rows have equal elements, and for $i > p$ it vanishes since the rank of $\| a_\beta^i \|$ is p. The coefficients A_σ^λ and a do not depend on the elements a_σ^i, a_1^i, ..., a_p^i, and so, writing

$$A_\sigma^\lambda / a = (-1)^{p-1} x_\sigma^\lambda,$$

we have

(2·3) $a_\sigma^i = x_\sigma^\lambda a_\lambda^i$, $(\sigma = p + 1, \ldots, m)$.

If $m > n$ we consider the points $a_\beta = (a_\beta^1, \ldots, a_\beta^n, 0, \ldots, 0)$, in the arithmetic space of m dimensions. We can then apply the above argument to obtain the relation (2·3), and the theorem is established for all values of m.

3. Linear sub-spaces.

If x_1, ..., x_k are k linearly independent points, the set of points linearly dependent on them will be called an *arithmetic linear k-space**, and the points x_1, ..., x_k will be said to *span* the linear k-space defined in this way. Thus a linear 1-space consists of the points whose components are proportional to those of a given point, and may conveniently be called an arithmetic straight line through the origin.

From the equations

(3·1) $x^i = t^\lambda x_\lambda^i$, $(\lambda = 1, \ldots, k)$

which define a linear k-space, X_k, it follows that to each point $(t^1, \ldots, t^k)$ of the arithmetic space of k dimensions corresponds a point of X_k, the points e_0 and e_1, ..., e_k corresponding to the origin and x_1, ..., x_k respectively. Moreover, to each point of X_k corresponds just one point in the k-dimensional arithmetic space. For if t_1 and t_2 are points in the latter corresponding to the same point in X_k, we have

$$t_1^\lambda x_\lambda^i = t_2^\lambda x_\lambda^i,$$

or $(t_2^\lambda - t_1^\lambda) x_\lambda^i = 0.$

But x_1, ..., x_k are independent and so $t_1 = t_2$.

Equations of the form (3·1), therefore, define not only a linear k-space, but a linear k-space which is in (1-1) correspondence with the arithmetic space of k dimensions. Such a correspondence is called a *parameterization* of the linear k-space.

All points linearly dependent on m points a_1, ..., a_m, in a linear k-space, X_k, are contained in X_k. For a_1, ..., a_m are given by equations of the form

$$a_\beta^i = t_\beta^\lambda x_\lambda^i,$$

---

* We shall define flat sub-spaces in general in § 7 below. The linear sub-spaces all contain the origin. They owe their importance to the fact that (with the notations explained in § 1) if a linear k-space contains two points x_1 and x_2, it also contains $x_1 + x_2$. We can express this by saying that linear k-spaces are closed under addition.

and any point linearly dependent on $a_1, \ldots, a_m$ is obviously dependent on $x_1, \ldots, x_k$. This may be called the transitive law for linear dependence.

If the points $x_1, \ldots, x_k$ span the linear k-space, X_k, there is no other linear k-space containing these points. For if $y_1, \ldots, y_k$ span a linear k-space, Y_k, containing $x_1, \ldots, x_k$, we have

(3·2) $$x^i_\lambda = t^\mu_\lambda y^i_\mu.$$

If the determinant $| t^\lambda_\mu |$ were zero there would be a relation of the form

$$s^\mu t^\lambda_\mu = 0,$$

in which $s^1, \ldots, s^k$ were not all zero. This would imply

$$s^\lambda x^i_\lambda = 0$$

and $x_1, \ldots, x_k$ would not be independent. For each value of i, therefore, the equations (3·2) can be solved by Cramer's rule to yield equations of the form

(3·3) $$y^i_\lambda = T^\mu_\lambda x^i_\mu.$$

From the transitive law for linear dependence, and from (3·2) it follows that each point of X_k lies in Y_k. Similarly it follows from (3·3) that each point of Y_k lies in X_k. They are, therefore, identical. We can express this by saying that a linear k-space is spanned by any set of k independent points contained in it, and it follows that a linear k-space does not contain a set of l independent points, where $l > k$. For the definition in §2 implies that any k points in a set of l independent points are themselves independent, and would therefore span any linear k-space containing the larger set.

The theorem of §2 can now be stated in the form: *If p is the rank of a matrix $\| x^i_\beta \|$, the points $x_1, \ldots, x_m$ are all contained in a linear p-space but not in a linear q-space, where $q < p$.*

4. Linear homogeneous transformations.

Any correspondence under which each point, x, in a set of points X corresponds to a unique point, y, is called a *single valued transformation* of X into Y, where Y is the set of points to which the points of X correspond. We may denote the transformation by

$$x \rightarrow y.$$

If no two distinct points in X correspond to the same point in Y, the transformation $x \rightarrow y$ will be called (1-1), or *non-singular*. If $x \rightarrow y$ is any non-singular transformation there exists a unique single-valued transformation, $y \rightarrow x$, called the *inverse* of $x \rightarrow y$.

A transformation which is given by equations of the form

$$(4\cdot1) \qquad\qquad y^i = a^i_j x^j$$

is said to be *linear* and *homogeneous*, and is non-singular if the determinant $a = |a^i_j|$ is not zero. For if $a \neq 0$ the equations $(4\cdot1)$, which are identical with $(1\cdot1)$, can be solved to obtain the inverse transformation

$$(4\cdot2) \qquad\qquad x^i = \alpha^i_j y^j,$$

where α^i_j is the normalized co-factor of a^j_i.

Any linear transformation $x \rightarrow y$, whether singular or not, will carry any point which is dependent upon a given set, $x_1, \ldots, x_k$, into a point which is dependent upon $y_1, \ldots, y_k$, where $x_\lambda \rightarrow y_\lambda$. For a point x, given by

$$(4\cdot3) \qquad\qquad x^i = t^\lambda x^i_\lambda,$$

goes into a point y, given by

$$(4\cdot4) \qquad\qquad y^i = a^i_j x^j = a^i_j t^\lambda x^j_\lambda = t^\lambda y^i_\lambda.$$

Not only the relation of linear dependence, therefore, but also the parameters $t^1, \ldots, t^k$, by which it is expressed, are unaltered by linear transformations.

The unit points are carried into the columns of the matrix $|| a^i_j ||$, and if p is the rank of the latter these columns will be contained in a linear p-space X_p, exactly p of them being independent. Hence the whole arithmetic space will be carried into X_p, and any two points, x_1 and x_2, such that

$$a^i_j (x^j_2 - x^j_1) = 0,$$

will be carried into the same point in X_p. The condition $a \neq 0$ is, therefore, not only sufficient, but also necessary in order that the transformation given by $(4\cdot1)$ shall be non-singular.

An independent set of points, $x_1, \ldots, x_k$, is carried by a non-singular linear homogeneous transformation*, $x \rightarrow y$, into an independent set $y_1, \ldots, y_k$. For if some of the latter were dependent upon the others we could apply the inverse transformation, $y \rightarrow x$, to show that the same was true of $x_1, \ldots, x_k$. Linear homogeneous transformations, therefore, carry linear k-spaces into linear k-spaces, and from the theorem in § 2 it follows that the matrices

$$|| x^i_a ||, \text{ and } || y^i_a || = || a^i_j x^j_a ||,$$

have the same rank.

---

* In this and the following chapters, all transformations are to be taken as non-singular unless the contrary is stated.

Cramer's rule depends upon the fact that if a_j^k are given, and if $a \neq 0$, there exists a matrix $||\alpha_j||$ which is uniquely determined by the condition

$$(4 \cdot 5) \qquad \delta_j^i = \alpha_k^i \, a_j^k .$$

If we regard the columns of $||a_j^k||$ as arithmetic points this states that there is just one linear homogeneous transformation which carries a given set of n independent points, $a_1, \dots, a_n$, into the unit points $e_1, \dots, e_n$.

As a corollary, we see that there is at least one linear homogeneous transformation which carries any set of k independent points, $a_1, \dots, a_k$, into the unit points $e_1, \dots, e_k$. For we can find $n-k$ points $a_{k+1}, \dots, a_n$ such that $a_1, \dots, a_n$ are linearly independent (if the determinant $|a_\mu^\lambda|$, $(\lambda, \mu = 1, \dots, k)$ is not zero we can take $a_{k+1} = e_{k+1}, \dots, a_n = e_n$) and there is just one transformation in which $a_i \to e_i$. A transformation which carries an independent set of points $a_1, \dots, a_k$ into $e_1, \dots, e_k$, will carry the linear k-space spanned by the former into that given by

$$(4 \cdot 6) \qquad \begin{cases} y^\lambda = t^\lambda, & (\lambda = 1, \dots, k) \\ y^\sigma = 0, & (\sigma = k + 1, \dots, n) \end{cases}$$

that is to say, into the set of all points satisfying the equations

$$y^{k+1} = 0, \ \dots, \ y^n = 0.$$

5. Homogeneous linear equations.

There is a linear homogeneous transformation

$$(5 \cdot 1) \qquad y^i = a_j^i \, x^j,$$

which carries a given linear k-space, X_k, into the linear k-space given by $(4 \cdot 6)$. It follows that X_k consists of those, and only those points, which satisfy the set of $n-k$ linear homogeneous equations

$$(5 \cdot 2) \qquad a_j^\sigma \, x^j = 0 \qquad (\sigma = k + 1, \dots, n).$$

Again, if $(5 \cdot 2)$ is any set of $n-k$ linear homogeneous equations in n variables, x, such that the matrix

$$||a_j^\sigma||$$

is of rank $n-k$, a transformation $(5 \cdot 1)$ can be found which carries the set of points satisfying $(5 \cdot 2)$ into the linear k-space $(4 \cdot 6)$. Since linear k-spaces are carried into linear k-spaces by linear homogeneous transformations, it follows that the solutions of $(5 \cdot 2)$ constitute a linear k-space.

If

$$(5\cdot3) \qquad b_j^{\alpha} x^j = 0, \qquad (\alpha = 1, \ldots, m)$$

is any set of linear homogeneous equations, such that the rank of the matrix $\| b_j^{\alpha} \|$ is r, there are r of these equations, which we may suppose to be

$$(5\cdot4) \qquad b_j^{\sigma} x^j = 0, \qquad (\sigma = 1, \ldots, r)$$

such that the matrix $\| b_j^{\sigma} \|$ is of rank r, and the remaining equations are linear combinations of $(5\cdot4)$. Hence the points satisfying $(5\cdot4)$ satisfy the full set of equations, and of course, any point satisfying $(5\cdot3)$ satisfies $(5\cdot4)$. By the last paragraph the points satisfying $(5\cdot4)$ constitute a linear $(n-r)$-space. *Therefore the solutions to a set of linear homogeneous equations constitute an $(n-r)$-space, where r is the rank of the matrix of the coefficients.*

Any set of points which span the $(n-r)$-space is called a *complete set* of solutions.

Taken with the description of the way in which a linear k-space is spanned by sets of k independent points, this summarizes the theory of linear homogeneous equations.

6. Translations.

A transformation given by equations of the form

$$(6\cdot1) \qquad y^i = x^i + a^i$$

is called a *translation*. It is obvious that translations are non-singular and that the inverse of a translation is a translation; also that if $x \to y$ and $y \to z$ are translations, the resultant transformation $x \to z$ is a translation; also that there is just one translation, namely that given by

$$y^i = x^i + y_0^i - x_0^i,$$

which carries a given point x_0 into a given point y_0.

7. Flat sub-spaces.

Any set of points which corresponds under a translation to a linear k-space will be called an *arithmetic flat k-space*. For $k = 0$, an arithmetic flat k-space is a single point; for $k = 1$ it is called an arithmetic straight line, for $k = 2$ a plane, and for $k = n - 1$ a hyperplane. If one of two flat k-spaces can be carried by a translation into the other, they are said to be *parallel*. From the transitive property of translations it follows that flat k-spaces are carried by translations into flat k-spaces, and that two flat k-spaces which are parallel to a third, are parallel to

each other. Through any point x_0 there is a flat k-space parallel to a
given linear k-space, X_k. This flat k-space is obtained from X_k by the
translation which carries the origin into x_0. Any flat k-space is parallel
to itself by this definition, since the identical transformation, which
leaves each point unaltered, is a special case of a translation.

If we apply the translation given by

$$(7\cdot1) \qquad\qquad y^i = x^i + y_0^i$$

to the linear k-space whose points satisfy $(n - k)$ linearly independent
linear homogeneous equations

$$(7\cdot2) \qquad\qquad a_j^\sigma x^j = 0, \qquad (\sigma = k+1, \ldots, n)$$

we find that any flat k-space is the set of points satisfying a set of
equations of the form

$$(7\cdot3) \qquad\qquad a_j^\sigma y^j = a_0^\sigma .$$

The constants a_0 are given by

$$a_0^\sigma = a_j^\sigma y_0^j,$$

and will be zero if, and only if, the point y_0 to which the origin is carried
by the translation is in the linear k-space $(7\cdot2)$. Similarly, if y_0 and y_0'
are any points in the flat k-space $(7\cdot3)$, the translation $y_0 \to y_0'$ carries
this flat k-space into itself. We recall that a flat k-space, Y_k, was
defined as the image of a linear k-space, X_k, under a translation $e_0 \to y_0$,
and it follows that Y_k is equally well defined as the image of X_k under
the translation $e_0 \to y_0'$, where y_0' is any point in Y_k. Therefore any
pair of flat k-spaces which are parallel to each other and have a common
point are identical. Hence there is one, and only one, flat k-space which
passes through a given point and is parallel to a given flat k-space.

Conversely, if we have a set of $n - k$ linear equations of the form
$(7\cdot3)$, such that the matrix of the coefficients on the left,

$$||a_j^\sigma||,$$

is of rank $(n - k)$, they are satisfied by a set of points which constitute
a k-space. To prove this, we first observe that we can transpose k of
the variables, y, to the right of $(7\cdot3)$, leaving on the left $n - k$ variables,
y, the determinant of whose coefficients is not zero. Then substitute
arbitrary values for the y's on the right, solve for the remaining ones
by Cramer's rule, and we have a set of values $y_0^1, \ldots, y_0^n$ which satisfy
$(7\cdot3)$. Now apply the translation $y \to x$ inverse to $(7\cdot1)$, and we find
that the set of points satisfying $(7\cdot3)$ is carried into the set of points

satisfying (7·2). In other words, the solutions of (7·3) satisfy the definition of a flat k-space.

By definition, a linear k-space is a set of points satisfying the equations (3·1). Applying the translation (7·1) we find that a flat k-space in general is a set of points satisfying equations of the form

$$(7·4) \qquad y^i = y^i_0 + t^\lambda (y^i_\lambda - y^i_0), \qquad (\lambda = 1, \dots, k).$$

Hence the theorem which we have just proved asserts that the solutions of a set of equations of the form (7·3) are given by equations of the form (7·4), and conversely, any set of points, y, given by equations of the type (7·4), are the solutions of a set of equations of the form (7·3).

The flat k-space (7·4) contains the points $y_0, y_1, \dots, y_k$. Substituting the right-hand side of (7·4) into the right-hand side of the formula for a homogeneous linear transformation, $y \to z$, it follows that (7·4) is carried by $y \to z$ into a flat k-space through the points $z_0, z_1, \dots, z_k$, where $y_\alpha \to z_\alpha$.

8. Non-homogeneous linear equations.

Consider a general set of linear equations

$$(8·1) \qquad a^\alpha_j x^j = a^\alpha_0, \qquad (\alpha = 1, \dots, m)$$

and let us refer to

$$\| a^\alpha_j \|$$

as the matrix of the coefficients, and to the matrix

$$\| a^\alpha_\sigma \|, \qquad (\sigma = 0, 1, \dots, n)$$

with the column a^α_0 adjoined, as the *augmented* matrix. Let r be the rank of the matrix of the coefficients and s the rank of the augmented matrix. Naturally, $s \geqq r$. We may assume without loss of generality that the rank of the matrix

$$\| a^\lambda_j \|, \qquad (\lambda = 1, \dots, r)$$

of the first r of the equations (8·1) is r, and therefore that the equations

$$(8·2) \qquad a^\lambda_j x^j = a^\lambda_0$$

are satisfied by all the points in some flat $(n-r)$-space, and only by those. If $s > r$ there is an equation

$$(8·3) \qquad a_j x^j = a_0$$

in the set (8·1), such that no relation of the form

$$(8·4) \qquad a_\sigma = p_\lambda a^\lambda_\sigma, \qquad (\lambda = 1, \dots, r; \; \sigma = 0, 1, \dots, n)$$

exists. But since the rank of the matrix $\|a_j^a\|$ is $r < s$, there is a relation of the form

$$a_t = p_\lambda a_i^\lambda.$$

If any solution, x, of (8·2) were to satisfy (8·3) we should have

$$\begin{cases} a_t = p_\lambda a_i^\lambda, \\ a_0 = a_j x^j = p_\lambda a_j^\lambda x^j \\ \quad = p_\lambda a_0^\lambda, \end{cases}$$

contradicting the hypothesis that no relation of the form (8·4) exists. Therefore *a set of linear equations is consistent if, and only if, the augmented matrix has the same rank as the matrix of the coefficients. If this rank is r the solutions constitute a flat $(n-r)$-space.*

9. Linear transformations.

The resultant of a linear homogeneous transformation (singular or non-singular) and a translation, is obviously given by equations of the form

(9·1) $y^i = a_j^i x^j + a^i,$

whichever is applied first. Such a transformation is called a *linear transformation*. Since linear homogeneous transformations and translations both carry flat sub-spaces into flat sub-spaces, it follows that any linear transformation has this property. By the application of a suitable translation, many properties of linear transformations in general may be deduced at once from the corresponding property of homogeneous linear transformations. For instance it follows that *the transformation* (9·1) *is non-singular if, and only if, $a \neq 0$,* and that *there is at least one linear transformation which carries $k+1$ given points $x_0, x_1, \ldots, x_k$ into $y_0, y_1, \ldots, y_k$ respectively, provided neither set of points is contained in a single flat $(k-1)$-space.* In particular, if $k = n$, there is just one such transformation.

The inverse of a non-singular linear transformation is obviously linear.

10. Affine theorems.

We are now in a position to prove by purely arithmetic methods a large number of theorems which are indicated by our geometric terminology. It will be sufficient to give a few examples.

There is one and only one arithmetic flat k-space containing $k+1$ given points, $y_0, y_1, \ldots, y_k$, which are not in the same flat $(k-1)$-space.

Let the translation $y_0 \rightarrow e_0$ carry the points $y_1, \ldots, y_k$ into $x_1, \ldots, x_k$ respectively. If $x_1, \ldots, x_k$ were contained in a linear $(k-1)$-space, X_{k-1}, the points $y_0, \ldots, y_k$ would be contained in a flat $(k-1)$-space parallel to X_{k-1}. Therefore $x_1, \ldots, x_k$ span a linear k-space X_k, and there is at least one flat k-space through $y_0, \ldots, y_k$, namely that which is parallel to X_k, and which is given by (7·4). If Y_k is any flat k-space through $y_0, \ldots, y_k$ it is carried by $y_0 \rightarrow e_0$ into a linear k-space which is parallel to Y_k, and which contains $x_1, \ldots, x_k$. But X_k is the only linear k-space spanned by $x_1, \ldots, x_k$, and therefore there is only one flat k-space through $y_0, y_1, \ldots, y_k$.

Since there is at least one linear transformation which carries $(k+1)$ given points $x_0, x_1, \ldots, x_k$ into $k+1$ given points $y_0, y_1, \ldots, y_k$, provided neither set is in a single flat $(k-1)$-space, and since linear transformations carry flat k-spaces into flat k-spaces, *there is at least one linear transformation which carries one of two flat k-spaces into the other.*

The straight line joining any two points of a flat k-space lies entirely in the k-space.

The flat k-space can be carried by a linear transformation into the linear k-space satisfying the equations

(10·1) $$x^{k+1} = 0, \ldots, x^n = 0.$$

The given line is carried into a line joining two points, x_1, x_2, satisfying (10·1). But all points of the line are given by the formula

$$x^i = x_1^i + t\,(x_2^i - x_1^i),$$

and they all satisfy the conditions (10·1), since x_1 and x_2 satisfy them. Hence the line lies entirely in the k-space.

It would be easy to continue with a long list of theorems of this class. They all refer to properties of figures which are unaltered by non-singular linear transformations, or *affine* transformations as they are sometimes called. We may therefore call the class of theorems which we have been illustrating, *affine theorems.*

11. We have seen that any linear transformation carries straight lines into straight lines. The converse theorem is also true, though not so obvious: *any non-singular transformation of the arithmetic space of n dimensions into itself* is linear if it carries straight lines into straight lines†.*

---

* By a transformation, $x \rightarrow y$, of the arithmetic space into itself we shall always mean a transformation of the whole space into the whole space. That is to say, each x corresponds to some y, and each y is the correspondent of some x.

† This is equivalent to one form of the fundamental theorem of projective geometry. The proof is essentially that given by Darboux, "Sur le théorème fondamental de la géométrie projective," *Math. Annalen,* Vol. 17 (1880), p. 55.

We shall prove this for $n = 2$ and leave the reader to generalize the theorem. Let T be any transformation of the arithmetic space into itself which carries straight lines into straight lines. By this we mean that each point on a given straight line l, corresponds under T to a point on some straight line m, and each point on m is the correspondent of some point on l. If x_0, x_1 and x_2 are non-collinear points, they are carried by T into non-collinear points, y_0, y_1 and y_2, say. For if y_2 were on the straight line $y_0 y_1$, it would be the correspondent under T of a point on $x_0 x_1$, contrary to the assumption that x_2 is not on $x_0 x_1$. Since y_0, y_1 and y_2 are non-collinear, a non-singular linear transformation, S, exists, which carries y_0, y_1 and y_2 back into x_0, x_1 and x_2, respectively.

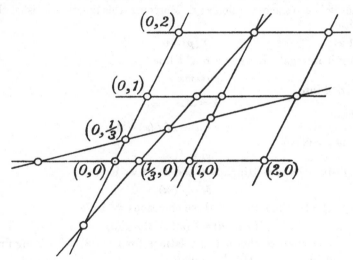

Fig. 1.

The resultant of T followed by S leaves x_0, x_1 and x_2 unaltered, and carries straight lines into straight lines. We shall show that such a transformation is the identity. It then follows that T is the inverse of a linear transformation, and hence is linear.

We may suppose, therefore, that the points e_0, e_1, e_2 are left unaltered. Since the transformation is non-singular, parallel straight lines are carried into parallel straight lines. For let two parallel lines l and m be carried into l' and m' respectively. If l' and m' are not parallel they will have a point x' in common, which will correspond to a point x in which l and m intersect. This contradicts the hypothesis that l and m are parallel. As indicated in the figure, therefore, it is easy to see, first that all points both of whose components are integers, and secondly that all points with rational components, are left unaltered. Since straight lines parallel to the axes are carried into straight lines parallel to the axes, the transformation is given by equations of the form

$$(11\cdot1) \qquad \begin{cases} X = f(x), \\ Y = \phi(y). \end{cases}$$

Since rational points are unaltered the line

$$y = x$$

is carried into itself. This means that

$$f(x) = \phi(x),$$

or that ϕ is the same function as f. Moreover, this function satisfies the condition

$$(11\cdot2) \qquad\qquad f(p) = p$$

when p is rational. Let any straight line

$$\hat{y} = mx + a$$

be carried into

$$Y = MX + A.$$

We have

$$f(mx + a) = Mf(x) + A.$$

Putting $x = 0$, we have

$$A = f(a),$$

since $f(0) = 0$; and putting $a = 0$ and $x = 1$ we have

$$M = f(m),$$

since $f(1) = 1$. From the last three equations we have

$$(11\cdot3) \qquad\qquad f(mx + a) = f(m)f(x) + f(a).$$

It is convenient to obtain two relations from $(11\cdot3)$ by putting first $m = 1$ and then $a = 0$. We thus obtain

$$(11\cdot4) \qquad\qquad \begin{aligned} &(a) \quad \begin{cases} f(x+y) = f(x) + f(y), \\ &(b) \quad f(xy) = f(x)f(y), \end{cases} \end{aligned}$$

where x and y are any real numbers*.

If h is any positive number it follows from $(11\cdot4b)$ that

$$f(h) = \{f(\sqrt{h})\}^2.$$

---

* These are the conditions obtained by Darboux on the assumption that the transformation of a straight line into itself given by

$$X = f(x)$$

carries harmonic sets into harmonic sets. In two dimensions it can be shown by elementary properties of the quadrilateral, that harmonic sets of points are carried into harmonic sets.

Since the transformation is non-singular, $f(h)$ vanishes only when $h = 0$, and from $(11\cdot4a)$ we have

$$(11\cdot5) \qquad f(x+h) - f(x) = f(h) > 0,$$

and therefore $f(x)$ is an increasing function.

Now let x be any number and let

$$y = f(x).$$

If $y > x$ there is a rational number p, such that $y > p > x$, and from $(11\cdot2)$ we have

$$0 < y - p = f(x) - f(p).$$

This contradicts $(11\cdot5)$, and a similar contradiction arises in assuming $y < x$. Hence $f(x) = x$ and the transformation T, defined by $(11\cdot1)$, is the identity.

12. The elementary distance function.

Another term borrowed by analysis from elementary geometry is *distance*, meaning the number

$$(12\cdot1) \qquad \delta(x, y) = \{(x^1 - y^1)^2 + \ldots + (x^n - y^n)^2\}^{\frac{1}{2}},$$

where the positive square root is taken. Thus for every two arithmetic points there is a positive number, their distance, which satisfies conditions familiar in elementary geometry. For example,

$$
\begin{aligned}
&(a) \quad &&\delta(x, x) = 0,\\
&(b) \quad &&\delta(x, y) \neq 0 \text{ if } x \neq y,\\
&(c) \quad &&\delta(x, y) = \delta(y, x),\\
&(d) \quad &&\delta(x, y) + \delta(y, z) \geqq \delta(x, z).
\end{aligned}
$$

The first three of these conditions are obvious. To prove the fourth, we observe first that if two points x, y are translated into two points x', y', then $\delta(x, y) = \delta(x', y')$, that is to say, mutual distances are unaltered by translations. The validity of formula (d) will therefore be unaltered if the points x, y, z are transformed by a translation which carries y into the origin. Thus (d) reduces to the inequality

$$(12\cdot2) \qquad \{\Sigma(x^i)^2\}^{\frac{1}{2}} + \{\Sigma(z^i)^2\}^{\frac{1}{2}} \geqq \{\Sigma(x^i - z^i)^2\}^{\frac{1}{2}},$$

which may be verified by simple algebra.

Equality occurs in $(12\cdot2)$ if, and only if, x and z are on the same straight line through the origin and if they are on different sides of the origin. That is to say equality occurs in (d) if, and only if, x, y and z are related by an equation of the form

$$\lambda(x^i - y^i) + \mu(z^i - y^i) = 0,$$

where λ and μ are not both zero, and have different signs. If this condition is satisfied, and if $x \neq y$ and $y \neq z$, y is said to lie *between* x and z.

It is not difficult to derive by purely arithmetic processes a series of theorems which make use of the geometric terminology which we have just indicated and which read like the theorems of elementary geometry. These theorems have to do with properties which are not altered by transformations of the arithmetic space which leave distances unaltered, i.e. transformations such that if $x \to \bar{x}$ and $y \to \bar{y}$, then

$$(12\cdot3) \qquad \Sigma\,(x^i - y^i)^2 = \Sigma\,(\bar{x}^i - \bar{y}^i)^2.$$

Since a point y is between two others, x and z, if, and only if,

$$\delta\,(x,\,y) + \delta\,(y,\,z) = \delta\,(x,\,z),$$

it follows that a transformation which leaves distances unaltered carries straight lines into straight lines. Hence by the theorem of § 11, the transformation is linear, that is to say it takes the form

$$(12\cdot4) \qquad \bar{x}^i = a^i_j\,x^j + a^i_0.$$

If we substitute these equations in ($12\cdot3$) we find that

$$(12\cdot5) \qquad a^i_j\,a^i_k = \delta_{jk}, \qquad (\delta_{jk} = \delta^j_k),$$

summation with respect to i being understood. Any linear homogeneous transformation which satisfies ($12\cdot5$) is called *orthogonal*, and its matrix an *orthogonal matrix*.

An affine transformation for which the determinant, a, is *positive* is said to be direct and one for which a is negative is said to be *opposite*. The direct orthogonal transformations are sometimes called *proper* and the opposite ones *improper*. If $n = 3$ the proper ones are called *rotations*. An affine transformation which leaves distances unaltered, i.e. which satisfies ($12\cdot5$), is called a *displacement* if it is direct and a *symmetry* if it is opposite.

CHAPTER II

GEOMETRIES, GROUPS AND COORDINATE SYSTEMS

1. A Geometry as a Mathematical Science.

During the nineteenth century mathematicians arrived at the notion that there are not one, but many geometries. This idea, which originated in attempts to prove Euclid's axiom of parallelism, reduced in the end to the following: Any mathematical science is a body of theorems deduced from a set of axioms. A geometry is a mathematical science. The question then arises why the name geometry is given to some mathematical sciences and not to others. It is likely that there is no definite answer* to this question, but that a branch of mathematics is called a geometry because the name seems good, on emotional and traditional grounds, to a sufficient number of competent people.

As the words are generally used at present a geometry is the theory of a space, and a space is a set of objects, usually called points, together with a set of relations in which these points are involved. A space, therefore, is not merely an abstract set of objects, but a set of objects with a definite system of properties†. These properties will be referred to as the *structure* of the space.

For example, the structure of what is called a metric space is defined by a function $\delta(P, Q)$, whose argument is a pair of points, and whose values are non-negative numbers. $\delta(P, Q)$ is called the *distance* between the points P and Q, and satisfies the conditions (cf. Chap. I, § 12):

$$\begin{cases} \delta(P, P) = 0, \quad \delta(P, Q) > 0 \text{ if } P \neq Q, \\ \delta(P, Q) = \delta(Q, P), \\ \delta(P, Q) + \delta(Q, R) \geq \delta(P, R). \end{cases}$$

* Any objective definition of geometry would probably include the whole of mathematics. Consider, for example, the spheres in a three-dimensional Euclidean metric space (§6) with a given point as centre. These spheres are in (1-1) correspondence with the positive real numbers, each sphere corresponding to its radius, and a negative number can be defined as a relation between two of them. Any theorem of analysis can, therefore, be translated into a theorem about these spheres, and it would be difficult to frame a definition of geometry which would discriminate among such theorems.

† We are using the word property in a very broad sense. A property may be a point, a set of points or a set of relations. The set of arithmetic straight lines, for example, is a property of the arithmetic space, so is the relation between parallel lines of not intersecting.

According to the definition which we have adopted, two distinct spaces may consist of the same set of points with different structures. For example, if we set up two distance functions in the arithmetic space of n dimensions we obtain two distinct metric spaces.

Two spaces U and V are said to be *equivalent* if there is a (1-1) correspondence, $P \to Q$, between the points, $[P]$, of U and the points, $[Q]$, of V, which sets up a (1-1) correspondence * between all the properties which constitute the structure of U, and those which constitute the structure of V. In this case $P \to Q$ may be said to carry the space U into the space V. The two spaces have the same geometry because every statement which can be made about the structure of U translates under the correspondence $P \to Q$ into a statement about the structure of V.

Two equivalent spaces may also be called *isomorphic* and a transformation of one into the other an *isomorphism*. In particular, if $U = V$ a transformation, $P \to Q$, which carries U into itself is called an *automorphism* of U.

2. Transformation groups.

There is an important class of geometries each of which can be regarded as the theory of a transformation group. A group is a set of elements which satisfy the following conditions:

(i)   With any ordered pair of elements a, b is associated an element c. We write

$$c = ab.$$

(ii)   The associative law is obeyed, that is

$$(ab)\, c = a\, (bc).$$

(iii)   There is an element i, called the unit element, such that

$$ai = ia = a.$$

---

* For example, if U and V are metric spaces, $P \to Q$ will carry U into V if, and only if, for each pair of points P and P' of U

$$\delta (P, P') = \delta (Q, Q'),$$

where $P \to Q$, $P' \to Q'$ and $\delta (Q, Q')$ is the distance function for V.

As another example consider the arithmetic space of one dimension. Let the points in the space be the real numbers and let the structure consist of the relations between any two numbers and their sum, and between any two numbers and their product. A transformation $x \to f(x)$ of the real numbers into themselves preserves this structure if, and only if, the function $f(x)$ satisfies the conditions (11·4) in Chap. I. Therefore the group of automorphisms (§ 3 below) of the space reduces to the identity. This is not true of the complex number system, for $z \to \bar{z}$ is an automorphism, where z and $\bar{z}$ are complex conjugate.

(iv) To each element, a, there corresponds an element, a^{-1}, which is said to be inverse to a, such that

$$a\,a^{-1} = a^{-1}a = i.$$

A correspondence under which each one of a given set of objects corresponds to an object in the set, and each object in the set is the correspondent of at least one object, is called a *transformation* of the set into itself. If this correspondence is (1-1), the transformation is said to be *non-singular*, and a non-singular transformation may also be described as a *permutation* of the objects among themselves. A set of permutations will obviously be a group under the following conditions:

(i)   The resultant of any two transformations in the set is also in the set.

(ii)   The inverse of each transformation in the set is also in the set. Such a set of permutations is called a *transformation group*.

The set of automorphisms of any space is obviously a transformation group. There are several examples of transformation groups in Chap. I:

(i)   The set of all translations (§ 6) is a group, the *translation group*.

(ii)   The set of all non-singular homogeneous linear transformations (§ 4) is a group, the linear homogeneous group or the *centred affine group*.

(iii) The set of all linear transformations (§ 9) is a group, the *affine group*.

(iv)   The set of all orthogonal transformations is a group, the *orthogonal group*.

(v)   The set of all displacements and symmetries is a group, the *Euclidean metric group*.

Each of these groups has a sub-group consisting of those of its transformations for which the determinant $a > 0$. Thus we have the direct linear homogeneous group, one direct affine group, the direct or proper orthogonal group, and the Euclidean displacement group.

3. Geometry and group-theory.

Let* $[P]$ be any set of objects which are permuted among themselves by a group G. The group G provides a method of classification by which any two figures (i.e. sets of points) in $[P]$ are in the same class if, and only if, there is a transformation of G which carries one into the other. Two figures in the same class are said to be *equivalent* or *congruent*.

* We use the symbol $[P]$ to stand for a set of which P is a typical member.

Because G is a group the relation of equivalence is transitive. The set of points $[P]$, with this classification for its structure, is a space whose group of automorphisms is G. Any property which is common to all the figures in one of the classes is said to be *invariant under G*, and the geometry of the space is often described as the study of properties which are invariant under G, or as the invariant theory of G.

As an example let the space consist of three objects A, B and C. The group shall be the set of even permutations,

$$\begin{pmatrix} ABC \\ ABC \end{pmatrix}, \quad \begin{pmatrix} ABC \\ BCA \end{pmatrix}, \quad \begin{pmatrix} ABC \\ CAB \end{pmatrix}.$$

The ordered triads fall into two classes,

$$(ABC, BCA, CAB) \text{ and } (BAC, ACB, CBA),$$

which may be taken to define positive and negative orientations of the space. The relation of likeness of orientation, i.e. of belonging to the same one of these classes, is a relation between two ordered triads which is left invariant by the group of this geometry.

4. An affine space.

Let the arithmetic space of n dimensions be the set of points, and the group the affine group. The space will be called an *affine** space and its geometry *affine geometry*.

Any two sets of points, one of which can be transformed into the other by an affine transformation, are said to be *affinely equivalent*. For example, it is proved in Chap. I that any two points, any two straight lines, and in general any two flat k-spaces are affinely equivalent. Also any two ordered sets of $n + 1$ points, neither of which is in a single flat $(n-1)$-space, are affinely equivalent.

The results of the first eleven sections in Chap. I may now be taken over as theorems in affine geometry. Moreover, a standard method of proof has been illustrated, namely the process of *normalization*. This consists in showing that a particularly simple representative of a given class of equivalent figures has a certain property, and then proving the invariance of the property under the group in question. A typical example is the proof of the third theorem in Chap. I, § 10, the linear k-space spanned by the first k unit points being chosen to represent the class of flat k-spaces. The method of normalization is not confined to geometries of the kind described in § 3, but can be used to advantage in the geometries discussed in the later chapters.

---

* Later on we shall refer to this as a flat affine space, as apart from generalized affine spaces.

The theorems of the class studied in Chap. I, § 10 are a sub-class of the theorems about the arithmetic space of n dimensions. In this sense the affine geometry of n dimensions is a sub-class of the theorems of analysis.

5. Affine spaces.

On the other hand, affine geometry is not merely a part of the theory of the arithmetic n-space. For example, consider the surface of a sphere in a space of three dimensions. The planes through a point P of this surface intersect it in a family of circles. If we refer to the circles through P, omitting P itself from each circle, as "straight lines," we have a family of points and straight lines with the same structure as the points and straight lines of a two-dimensional affine space.

The two-dimensional affine space defined in this way is in (1-1) correspondence with the arithmetic space of two dimensions in such a way that the straight lines of the former correspond to the arithmetic straight lines. According to the definition in § 1, the two spaces are isomorphic or equivalent, and have the same geometry.

There are, in fact, an infinity of n-dimensional affine spaces, all equivalent and having the same geometry. One of these is the arithmetic space of n dimensions plus the affine group. Affine geometry is the theory of what is alike in all these affine spaces.

6. Euclidean metric spaces.

As another example of a geometry let the space be the arithmetic space of n dimensions, and the group the Euclidean metric group. The geometry is then called the *Euclidean metric* geometry. Any point is "congruent," i.e. equivalent under this group, to any other point. So is any k-space congruent to any other k-space. But two pairs of points, xy and $x'y'$, are congruent if, and only if, $\delta(x, y) = \delta(x', y')$.

It is characteristic of this geometry that there is an invariant unit of distance, for the pairs of points x, y such that $\delta(x, y) = 1$ are distinguished by this fact from all other pairs of points.

7. Euclidean geometry.

The geometry described in the last section must not be confused with the Euclidean geometry. A special case of an n-dimensional Euclidean space is the arithmetic n-space with the group consisting of those affine transformations,

$$y^i = a_j^i x^j + a_0^i$$

for which

$$a_j^i a_k^i = \rho \delta_{jk}, \qquad (\rho > 0)$$

This group is called the *similarity group* or the *Euclidean group*. Two figures equivalent under it are said to be *similar*. Two figures equivalent under the Euclidean metric group (the sub-group obtained by requiring $\rho = 1$) are said to be *congruent*.

While two pairs of points are congruent only when equidistant, any two pairs of points are similar. But two triads of points are similar if, and only if, certain conditions are satisfied.

Among the special Euclidean spaces of three dimensions is the space of the external world as it appears in pre-relativity physical theories. Two figures are similar if they have "the same shape" and they are congruent if, in addition to being similar, they have the "same size." There is no uniquely determined unit of distance.

8. Coordinate systems.

A geometry need not necessarily be specified, as we have done in the three instances above, by giving a special example of a space with its structure defined by means of a transformation group. It can also be specified by means of a set of axioms, i.e. a set of statements from which all the theorems of the geometry are deducible. This has often been done for the Euclidean and affine geometries. We propose now to give yet another set of axioms, the excuse for which is that they are stated in terms of coordinate systems which are to be fundamental in our investigation of a much more extensive class of geometries. The axioms will presuppose the theory of the arithmetic space, and will describe any one of the spaces of a given geometry by means of its similarity to the arithmetic space.

When we say "point" we mean simply one of the completely undefined elements of the space which is to be characterized by our axioms. We shall sometimes distinguish this space from the arithmetic space by calling it a "geometric space." With a like purpose we shall sometimes refer to its points as "geometric points."

A *coordinate system* is a correspondence, $P \to x$, between a set of points $[P]$ and a set of arithmetic points $[x]$. Each geometric point P is said to be *represented* by each* arithmetic point to which it corresponds under $P \to x$. Any arithmetic point, x, which represents P is called an *image* of P in $P \to x$, and P is called an image of x. The numbers, $x^1, \ldots, x^n$, which constitute any image of P, are called *coordinates* of P.

An example of a coordinate system is a parameterization $x \to t$ of an

---

* The transformation $P \to x$ need not be (1-1) but may carry each P into a set of arithmetic points, as, for example, in a homogeneous coordinate system (cf. § 14 below).

arithmetic flat k-space, given by equations of the form (3·1) in Chap. I, §3. Here the arithmetic points $[x]$ play the part of $[P]$ and the points in the arithmetic space of k dimensions play the part of $[x]$. The term *parameterization* is often used instead of coordinate system and a set of points $[P]$ is said to be *parameterized* when referred to a coordinate system. A set of points is a different object from the same set of points parameterized in some way or other.

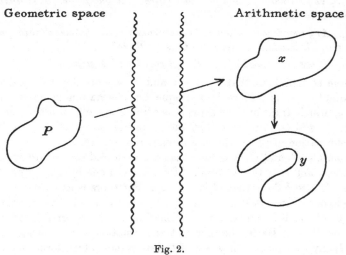

Fig. 2.

If $[y]$ is a set of arithmetic points in a correspondence* with $[x]$, the resultant of the transformation $P \to x$, followed by the transformation $x \to y$, is another coordinate system, $P \to y$. We shall call $x \to y$ a *transformation of coordinates* from $P \to x$ to the coordinate system $P \to y$. A transformation of coordinates is given by equations of the form

$$y^i = y^i(x).$$

The inverse transformation $y \to x$, given by

$$x^i = x^i(y),$$

is a transformation of coordinates from $P \to y$ to the coordinate system $P \to x$.

We shall use the name *coordinate geometry* for the theory of a space which can be completely described by means of coordinate systems.

* The transformation $x \to y$ need not be (1-1), e.g. when $P \to x$ is a homogeneous, and $P \to y$ a non-homogeneous coordinate system.

9. A class of coordinate geometries.

Let G be a set of transformations of arithmetic points into arithmetic points, and let us write down the following axioms, in which the undefined elements are *points* and *preferred coordinate systems*.

G_1. *Each preferred coordinate system is a (1-1) transformation of the space into the arithmetic space of n dimensions.*

G_2. *Any transformation of coordinates from one preferred coordinate system to another belongs to G.*

G_3. *Any coordinate system obtained from a preferred coordinate system by a transformation belonging to G is preferred.*

G_4. *There is at least one preferred coordinate system.*

From G_1 it follows that if $P \to x$ and $P \to y$ are any two preferred coordinate systems, there is a unique transformation of coordinates, $x \to y$, which transforms the first coordinate system into the second. From G_1 and G_3 it follows that each transformation of G is a (1-1) transformation of the arithmetic n-space into itself*.

From G_2, G_3 and G_4 it follows that the preferred coordinate systems are those, and only those, obtained from a given one by transformations of G. It also follows that G is a group. For there is at least one preferred coordinate system, $P \to x$, by G_4, and by G_3 any transformation, $x \to y$, of G determines a transformation to a preferred coordinate system, $P \to y$. By G_2 the inverse transformation $y \to x$ belongs to G. By G_3 any transformation $y \to z$ of G determines a transformation from $P \to y$ to a preferred coordinate system, $P \to z$. The transformation of coordinates from $P \to x$ to $P \to z$ is the resultant of $x \to y$ followed by $y \to z$, and belongs to G by G_2. Hence G satisfies the conditions for a transformation group given in §2.

The theory deduced from the axioms G is called the *geometry of the group G*.

Two spaces satisfying the axioms G, for the same group G, are obviously equivalent. For their points can be put in a (1-1) correspondence in such a way that to each preferred coordinate system of the one corresponds a preferred coordinate system of the other.

The arithmetic space of n dimensions plus the group G is an example of a space satisfying the axioms G. For if the transformations of G are

---

* This condition excludes, among others, the group of linear fractional transformations

$$y^i = \frac{a_j^i x^j + a^i}{a_j x^j + a}.$$

taken as preferred coordinate systems, the identity being the coordinate system in which each point corresponds to itself, the axioms are satisfied. Therefore the geometry of the group G according to § 3, is the same as the geometry of G according to this section. Indeed, the axioms G constitute an analysis of the following statement: The space is equivalent to the arithmetic space plus the group G.

If G is the affine group, for example, any space satisfying the axioms G is an affine space. The preferred coordinate systems are called *cartesian coordinate systems*. The image, in a cartesian coordinate system of an arithmetic straight line is called a straight line, and from Chap. I, § 11 it follows that, for $n > 1$, cartesian coordinate systems are those, and only those, transformations of an affine space into the arithmetic space which carry straight lines into arithmetic straight lines.

Likewise if G is the Euclidean metric group, $G_1, \ldots, G_4$ are a set of axioms for the Euclidean metric geometry, and if G is the similarity group they are a set of axioms for Euclidean geometry. The preferred coordinate systems for a Euclidean metric space, or for a Euclidean space, are called *rectangular cartesian coordinate systems*. The set of rectangular cartesian coordinate systems for a Euclidean space is a sub-set of the cartesian coordinate systems for an affine space, and a Euclidean space is an affine space with an additional element of structure, namely the class of properties defined by this smaller class of preferred coordinate systems. In general, if G' is a sub-group of G, i.e. a sub-set of the transformations belonging to G which is itself a group, a space satisfying the axioms G' is a space satisfying the axioms G, with an additional element of structure. The smaller the group, the more complicated the space *.

10. Centred affine geometry.

A space which satisfies the axioms G, where G is the group of linear homogeneous transformations, is called a *centred affine space* of n dimensions. The point which corresponds to the origin in any one, and therefore in all, preferred coordinate systems, is called the *centre* of the space.

A centred affine space, A_n^0, may be obtained by selecting an arbitrary coordinate system K_0 for an affine space, A_n, and taking it as a preferred

---

* The extreme cases arise when G is the group of all permutations of the arithmetic space, and when G is the identity. In the first case the space has no structure beyond its cardinal number, and in the second it has all the properties of the arithmetic space.

coordinate system for A_n^0. This amounts to selecting an arbitrary point in A_n as centre for A_n^0. The centred affine space having P_0 as its centre will be carried into that which has any other point $\overline{P}_0$ as its centre by any affine transformation which carries P_0 into $\overline{P}_0$.

The points of a centred affine space may also be referred to as vectors, and the space as a vector space. Addition and other operations with vectors can either be defined analytically by means of preferred coordinate systems and the arithmetical definitions in Chap. I, §1, or by means of geometrical constructions involving straight lines and parallelism. The coordinates of a vector in a preferred coordinate system are referred to as the *components* of the vector.

Let us say that two ordered point pairs, AB and $A'B'$, of an affine space are *equipollent* provided that the translation (Chap. I, §6) which carries A to A' also carries B to B'. Thus each ordered point pair determines a class of all ordered point pairs which are equipollent with it. There is a (1-1) correspondence between the points (vectors) of a centred affine space and these classes of equipollent ordered point pairs of an affine space. Namely, let O be any point of the affine space and let any point P, of the centred affine space with O as centre, correspond to the class of equipollent point pairs which includes OP. The class of point pairs AB, in which $A = B$, corresponds to O, the "null vector." Thus the theory of vectors in an affine space, where vectors are defined as objects not having a definite localization, but such that whenever an "initial point" and a vector are given a definite "terminal point" is determined, is a centred affine geometry.

Similarly there are centred Euclidean and Euclidean metric spaces, which satisfy the axioms G, where G is the group of homogeneous similarity transformations, and the orthogonal group respectively. In a centred Euclidean metric space each vector has a length, which is its distance from the centre. Any two vectors x and y have a scalar product, given in preferred coordinates by $x^i y^i$, which measures the angle between them, multiplied by the product of their lengths. They also have a vector product, and so on. It is this geometry which appears in most elementary books on vector analysis.

11. Oriented Spaces.

A space which satisfies the axioms $G_1, ..., G_4$, when G is the group of direct affine transformations,

$$y^i = a_j^i x^j + a_0^i, \qquad (a > 0)$$

is called an *oriented affine space*. It is an affine space plus an additional element of structure determined by a class of preferred coordinate systems within the class of cartesian coordinate systems. The latter fall into two classes, consisting of those obtainable from a given cartesian coordinate system by direct and by opposite transformations respectively. The coordinate systems in each class are obviously the preferred coordinate systems for an oriented affine space.

Thus any affine space uniquely determines two oriented affine spaces which are affinely equivalent to each other. One of these may be arbitrarily designated as *positively oriented* (or right-handed) and the other as *negatively oriented* (or left-handed). This process of naming the oriented affine spaces is called *orienting* the affine space.

An affine space may be oriented by choosing an arbitrary cartesian coordinate system $P \to x$, and specifying that oriented affine space as positively oriented for which $P \to x$ is a preferred coordinate system. All preferred coordinate systems of this oriented affine space are then said to be positively oriented. The coordinate system $P \to x$ may be specified by choosing $n + 1$ points, $P_0, P_1, ..., P_n$ which are not in the same hyperplane, and requiring the coordinate system to be that cartesian coordinate system in which $P_0, P_1, ..., P_n$ correspond to the arithmetic points $e_0, e_1, ..., e_n$ respectively.

The ordered sets of $n + 1$ points which are not in the same hyperplane fall into two classes, which we call *sense-classes*, according as the coordinate system in which $P_\alpha \to e_\alpha$ is positively or negatively oriented. The ordered sets of points of the first class are said to have *positive sense*, those of the other class, *negative sense*.

With obvious modifications this same discussion can be made for Euclidean and centred affine spaces. In the latter case a positive sense is fixed by specifying as positively oriented the preferred coordinate system in which an arbitrarily chosen set of n linearly independent vectors, $P_1, ..., P_n$, correspond to the arithmetic points $e_1, ..., e_n$ respectively. An ordered set of n independent vectors, represented in a positively oriented coordinate system by $x_1, ..., x_n$ respectively, obviously has positive or negative sense according as the determinant

$$| x_\alpha^i |$$

is positive or negative.

The transformation of coordinates given by

$$y^i = x^{\alpha i}$$

is direct if

(11·1) $$e_{a_1...a_n} = +1$$

and opposite if

$$(11\cdot2) \qquad e_{a_1 \ldots a_n} = -1.$$

It follows that the ordered sets of vectors, $P_1, \ldots, P_n$ and $P_{a_1}, \ldots, P_{a_n}$, have the same sense if $(11\cdot1)$ holds and opposite senses if $(11\cdot2)$ holds·

12. Oriented curves.

The whole theory of orientation is a generalization of the orientation of a straight line, and this is a translation into mathematics of the physical observation that a straight line joining two points, A and B, can be described by a particle moving in two ways, from A to B and from B to A.

More generally, let AB be any arc of a continuous curve which is given parametrically by

$$(12\cdot1) \qquad x^i = x^i(t), \qquad (t_0 \leqq t \leqq t_1)$$

where $A \to t_0$, $B \to t_1$, and the functions $x^i(t)$ are continuous. Analogous to the preferred coordinate systems for an oriented affine space, there is a class of parameterizations for the arc AB, any one of which is related to the parameterization $(12\cdot1)$ by a transformation of the form

$$s = s(t),$$

where $s(t)$ is a continuous function of t which increases steadily from t_0 to t_1. There is another class obtained from $(12\cdot1)$ by transformations, $t \to s$, where $s(t)$ is a continuous decreasing function. The arc AB associated with either of these classes of parameterizations is called an oriented arc. Thus any arc joining A to B determines two oriented arcs which we may denote by AB and by BA respectively.

13. Affine parameterizations.

In cartesian coordinates for an affine n-space, a straight line is given by linear parametric equations

$$(13\cdot1) \qquad x^i = x_0^i + t(x_1^i - x_0^i).$$

The parameterizations given by equations of this form are called *affine parameterizations*. They are obviously those, and only those, related to a given affine parameterization by linear transformations of parameter. Therefore a straight line, with affine parameterizations for cartesian coordinate systems, satisfies the axioms G for an affine space of one dimension*. Hence the straight line determines two oriented, or directed straight lines, which may be called $x_0 x_1$ and $x_1 x_0$, respectively, the

---

* Affine spaces of one dimension are Euclidean spaces, for the affine group in a 1-space is the same as the Euclidean group.

former being the one for which (13·1) is a preferred coordinate system.

This can be generalized to any flat k-space. Such a space has a family of affine parameterizations, and determines two oriented affine spaces. Any ordered set of $k + 1$ points x_0, x_1, ..., x_k, which are not in the same flat $(k - 1)$-space, determine an oriented k-space, namely the one which has the parameterization

$$x^i = x_0^i + t^\lambda (x_\lambda^i - x_0^i), \qquad (\lambda = 1, ..., k)$$

for a preferred coordinate system.

14. Projective and conformal geometry.

Projective and conformal n-spaces may also be described by axioms stated in terms of preferred coordinate systems. But in neither case can the whole space be represented in a preferred coordinate system by the arithmetic n-space. Therefore the axioms G must be modified. One way of doing this is to introduce homogeneous coordinate systems in which a point P, if it corresponds to an arithmetic point $(Z^1, ..., Z^m)$, also corresponds to the point $(\lambda Z^1, ..., \lambda Z^m)$, where λ is any non-zero factor. No point corresponds to the origin of the arithmetic space. Thus each geometric point is represented in homogeneous coordinates by each point, except the origin, on an arithmetic straight line through the origin.

We give a set of axioms for n-dimensional projective geometry in terms of preferred homogeneous coordinate systems.

P_1. *In a preferred homogeneous coordinate system each point is represented by at least one arithmetic point in the arithmetic space of $n + 1$ dimensions, and each arithmetic point other than the origin represents just one point.*

P_2. *Two arithmetic points represent the same point if, and only if, they lie on the same arithmetic straight line through the origin.*

P_3. *Any preferred coordinate system can be transformed into any other by a linear homogeneous transformation.*

P_4. *Any homogeneous coordinate system obtained from a preferred coordinate system by a linear homogeneous transformation is a preferred coordinate system.*

P_5. *There is at least one preferred coordinate system.*

From the axioms P_1 and P_2 it follows that each preferred coordinate system is a (1-1) correspondence between the points in the projective space and the arithmetic straight lines through the origin of the arithmetic $(n + 1)$-space.

A set of axioms for conformal geometry is obtained by modifying $P_1, \ldots, P_5$ so that each preferred coordinate system for a conformal n-space is a (1-1) correspondence between points in the conformal space and the generators of the hypercone

$$Z^i Z^i - 2Z^0 Z^{n+1} = 0$$

in the arithmetic $(n + 2)$-space.

The group of automorphisms of a projective space is called the n-dimensional projective group, and the group of a conformal n-space is called the n-dimensional conformal group. Projective and conformal spaces belong to the class described in § 3, for their structures are defined by the projective and conformal groups respectively.

15. Point transformations. Automorphisms.

Let $P \to \bar{P}$ be any transformation which permutes among themselves the points of a space satisfying the axioms given in § 9. To each preferred coordinate system $P \to x$, corresponds a coordinate system $\bar{P} \to x$, which is the resultant of the inverse transformation $\bar{P} \to P$,

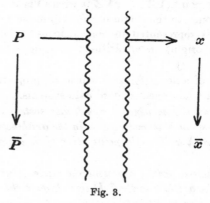

Fig. 3.

followed by $P \to x$. The transformation $P \to \bar{P}$ will be said to carry the coordinate system $P \to x$ into the coordinate system $\bar{P} \to x$, and will be an automorphism if, and only if, each preferred coordinate system is carried into a preferred coordinate system. For the structure of the space is determined by the totality of preferred coordinate systems.

In any preferred coordinate system, which we now denote by K instead of by $P \to x$, the transformation $P \to \bar{P}$ determines, and is determined by, a transformation $x \to \bar{x}$ in the arithmetic space, which carries the image in K of each point P into the image in K of the corresponding $\bar{P}$ (see diagram). The transformation $P \to \bar{P}$ is said to be *represented* in K by $x \to \bar{x}$. The coordinate system $\bar{K}$, into which K is carried by $P \to \bar{P}$, is obtained from K by the inverse transformation $\bar{x} \to x$ in the arithmetic space. Therefore the necessary and sufficient condition that $P \to \bar{P}$ be an automorphism is that it is represented in K by a transformation belonging to G. In this case it will obviously be represented in all preferred coordinate systems by transformations belonging to G. Therefore the group of automorphisms is represented in each preferred coordinate system by the group G, and these two groups are simply isomorphic. Any preferred coordinate system is an isomorphism which carries the group of automorphisms into G.

16. Changing views of geometry.

The concept of a mathematical science as a body of theorems deduced from axioms seems to have been clearly understood by Aristotle if not by earlier Greek scientists. This question is discussed by H. Scholz, "Die Axiomatik der Alten," in *Blätter für Deutsche Philosophie*, Vol. 4 (1930), p. 259.

For the Greeks there seems to have been only one space and this was not necessarily a collection of points, but rather a locus in which bodies could be moved about and compared with each other. The fundamental relation between bodies was congruence or superposability.

It was after the development of analytic geometry that space came to be regarded as a collection of points. With the advent of the non-Euclidean geometries it became accepted that there are many geometries. But the space was still a locus in which figures were to be compared, the central idea being that of the group of congruent transformations of the space into itself. Thus a geometry came to be regarded as the invariant theory of a transformation group. This point of view was formulated in the Erlanger Programm in 1870 (F. Klein, *Gesammelte Mathematische Abhandlungen*, Berlin, 1921, Vol. 1, p. 460).

The idea of a transformation group synthesized and generalized all previous concepts of motion and congruence. It also supplied a principle of classification by which it is possible to get a bird's-eye view of the relations between a large number of important geometries as, for example, all those discussed in the present chapter. For a more extensive discussion of this method of classification the reader may consult the second volume of Veblen and Young, *Projective Geometry*, New York, 1917, especially Chap. III.

The transition to the view of geometry as a special case of a mathematical science, the latter being a body of theorems deduced from axioms, was being made at about the beginning of the present century, when a great many different sets of axioms were stated and studied for a great variety of geometries by Pasch, Peano, Hilbert, Pieri, E. H. Moore, R. L. Moore, and others. On sets of axioms for affine and Euclidean geometry see A. N. Whitehead, *Axioms of Descriptive Geometry*, Cambridge tract, no. 5, Cambridge, 1907, and H. G. Forder, *The Foundations of Euclidean Geometry*, Cambridge, 1927. The word affine is not used in these books, but affine geometry is defined by the axioms of Euclidean geometry which do not refer to congruence.

The general concept of a mathematical science did not disturb the Erlanger Programm in any way. For a geometry could be described as a mathematical science which is the theory of a transformation group. But long before the Erlanger Programm had been formulated there were geometries in existence which did not properly fall within its categories, namely the Riemannian geometries. We need not define a Riemannian space here, but merely remark that there are Riemannian spaces which are metric spaces (in the sense given these words in § 1, above) whose groups of automorphisms reduce to the identity. Such a geometry obviously cannot be characterized by its group. A Riemannian space, however, has a structure in the sense explained at the beginning of the chapter. We may speak of the length of any smooth curve, and there is a system of curves (the geodesics) which have properties analogous to those of the straight lines.

The notion of a space as a set of points with a structure may be said to originate with Riemann's *Habilitationsschrift* (1854) (*Gesammelte Mathematische Werke*, Leipzig, 1876, p. 254), and has become well understood since its application to Einstein's General Relativity. For in this theory physical space is no longer a locus in which objects are moved about, but space-time is itself the only object studied in a complete geometry. There is no such thing as a body in space, but matter is an aspect of the space-time structure.

At the present time there are being actively studied, both for physical and for mathematical purposes, a large class of spaces which are generalizations of the Riemannian spaces. Like the Riemannian geometries they are, in general, outside the categories of the Erlanger Programm, but they all make use of coordinates and of differential calculus. This implies, as is shown in Chap. v below, that they make use of the "tangent spaces" or "spaces of differentials" which are centred affine spaces. In many cases they make use of other transformation groups in the tangent spaces than the centred affine group (see Chap. v, § 15, and Chap. vii, § 6).

There is, therefore, a strong tendency among contemporary geometers to seek a generalization of the Erlanger Programm which can replace it as a definition of geometry by means of the group concept. On this subject the reader is referred to J. A. Schouten, "Erlanger Programm und Uebertragungslehre. Neue Gesichtspunkt zur Grundlegung der Geometrie," *Rendiconti del Circolo Mat. di Palermo*, Vol. 50 (1926), p. 142, and to E. Cartan, "Les Récentes Généralisations de la Notion d'Espaces," *Bulletin des Sciences Math.* Vol. 48 (1924), p. 294; also to a paper by the same author called "Rapport sur le Mémoire de J. A. Schouten intitulé 'Erlanger Programm...'," *Bull. de la Soc. Phys. Math. de Kazan*, Series 3, Vol. 2 (1927), p. 71. The more general point of view that a geometry is the theory of a space with an invariant (the same thing as a geometric object in the sense of Chap. iii below) is set forth by O. Veblen, "Differential Invariants and Geometry," *Atti del Congresso Internazionale dei Matematici*, Bologna (1928), Vol. 1, p. 181.

ALLOWABLE COORDINATES

1. Functions of class u.

In the last chapter we have dealt with certain limited classes of coordinate systems, all of them such that the transformations of coordinates are linear. In general it is desirable to use a much larger class of coordinate systems, so that the transformations of coordinates shall be as general as they can be without destroying the significance of the analytic expressions which are to be used. The theory of the transformations which we shall. use depends upon the implicit function theorem in much the same way that the algebra contained in Chap. I depends upon Cramer's rule for solving linear equations. We shall need a few definitions.

A set of points in the arithmetic space of n dimensions given by inequalities of the form

$$(1\cdot1) \qquad |x^i - x_0^i| < \delta,$$

where $\delta > 0$, will be called a *box*, and x_0 will be called its *centre*. A set of points, $[x]$, is called a *region* if, and only if, each x is the centre of a box which is contained in $[x]$. Thus any box is a region, and so is the arithmetic space. Moreover, the set of points, X, common to two regions, X_1 and X_2, is a region. For if x_0 is any point in X, there is a box, given by

$$|x^i - x_0^i| < \delta_1,$$

which is contained in X_1, and a box given by

$$|x^i - x_0^i| < \delta_2,$$

which is contained in X_2. Therefore the box given by

$$|x^i - x_0^i| < \delta,$$

where δ is the smaller of δ_1 and δ_2, is contained in X.

A function* $F(x^1, \ldots, x^n)$, defined for all points in a region X, is said to be of *class u*, if it and its derivatives of order less than or equal to u exist and are continuous at each point of X. Here u can be any positive integer, and a function will be described as belonging to the class ∞ if all its derivatives exist. Continuous functions will be described as be-

---

* Unless otherwise stated it is to be assumed that all functions to which we refer are single valued.

longing to the class 0. A function, defined in a region X, is said to be *analytic* if it can be expanded in a power series about each point x_0 of X, the power series being convergent for all points in some box with centre x_0. Analytic functions will be described as belonging to the class ω. Of course a function which is of class u in a region X, is of class u in any region contained in X, and is also of class u' if $u' < u$.

2. The implicit function theorem.

For $u > 0$ the analogue of Cramer's rule is the implicit function theorem*, which we shall state without proof.

Let

(2·1)
$$\begin{cases} F^1 (x^1, \ldots, x^n; y^1, \ldots, y^m), \\ \vdots \\ F^n (x^1, \ldots, x^n; y^1, \ldots, y^m) \end{cases}$$

be n functions which are of class $u > 0$ for values of x and y in regions X and Y in the arithmetic spaces of n and m dimensions, respectively. Let there be a point x_0 in X and a point y_0 in Y such that

$$F^a (x_0; y_0) = 0,$$

and let the Jacobian

$$\frac{\partial (F^1, \ldots, F^n)}{\partial (x^1, \ldots, x^n)}$$

be different from zero for $x = x_0$ and $y = y_0$. The theorem states that the equations

(2·2) $F^a (x; y) = 0$

admit a unique set of solutions

$$x^i = x^i (y),$$

where $x^i (y)$ are functions of class u in some box contained in Y having y_0 as its centre, and where

$$x_0^i = x^i (y_0).$$

3. Transformations of class u.

Let $y^1 (x), \ldots, y^n (x)$ be n functions of class $u > 0$ in some region X. By means of the equations

(3·1) $y^i = y^i (x)$

each point x is made to correspond to a point y, and X will correspond to a set of points Y. Such a correspondence will be called a

---

* E. Goursat, *Cours d'analyse mathématique*, Paris, 1923, Vol. 1. The theorem is proved for $u = 0, 1, \ldots, \infty$ in Chap. III, §38, and for analytic functions in Chap. IX, § 185. (English translation, Goursat-Hedrick, Boston, 1904, Chap. III, § 25 and Chap. IX, § 188.)

transformation of class u which carries X into Y. We may also refer to $x \to y$, defined by (3·1), as operating on X. Since the functions $y^1(x), \ldots, y^n(x)$ are of class u in any region X', contained in X, the equations (3·1) will define a transformation of class u operating on any region contained in X.

Let the Jacobian $\left| \dfrac{\partial y}{\partial x} \right|$ be different from zero at each point of X. The implicit function theorem can then be applied to the equations

$$(3·2) \qquad\qquad y^i - y^i(x) = 0.$$

If we write

$$y_0^i = y^i(x_0),$$

where x_0 is any point in X, the equations (3·2) will admit a set of solutions

$$(3·3) \qquad\qquad x^i = x^i(y),$$

where $x^i(y)$ belong to the class u in some box, B_y, with y_0 as centre. Then B_y, being the image under (3·1) of a sub-set of X, is contained in Y. Hence Y is a region.

If the transformation (3·1) is a (1-1) correspondence of X to Y, then it has an inverse, $y \to x$, defined by equations of the form

$$(3·4) \qquad\qquad x^i = x^i(y).$$

Let y_0 be any point in Y. We have just shown that $x^i(y)$ are of class u in some box with y_0 as centre, and they are, therefore, of class u in Y.

A transformation of class $u > 0$, operating on a region X, will be called *regular* if, and only if, it is (1-1), and if its Jacobian does not vanish at any point* of X.

Unless otherwise stated it is to be assumed that all transformations to which we refer are regular. If $x \to y$ and $y \to z$ are two transformations of class u, which carry the region X into Y and Y into Z, respectively, the resultant, $x \to z$, which carries X into Z, will also be of class u. For $x \to z$ will be given by

$$z^i = z^i \{y(x)\},$$

---

* These two conditions are independent, as may be seen from the example $(x, y) \to (u, v)$, where $u = e^x \cos y$, $v = e^x \sin y$, and the example $y = x^3$. In the first example $\dfrac{\partial (u, v)}{\partial (x, y)} = e^{2x}$ and never vanishes. But the transformation $(x, y) \to (u, v)$, defined for all values of (x, y), is not (1-1). In the second example the transformation $x \to y$ is (1-1), but $\dfrac{dy}{dx}$ vanishes at the origin.

and the derivatives, $\dfrac{\partial^s z^i}{\partial x^{j_1} \dots \partial x^{j_s}}$, may be calculated by the ordinary rules of differential calculus, while if $u = \omega$, z^i can be expanded in a power series (Goursat, Chap. IX) about each point in X. The Jacobian of $x \rightarrow z$ is given by

$$(3\cdot5) \qquad\qquad \left| \frac{\partial z}{\partial x} \right| = \left| \frac{\partial z}{\partial y} \right| \left| \frac{\partial y}{\partial x} \right|,$$

and it follows that $x \rightarrow z$ is regular if $x \rightarrow y$ and $y \rightarrow z$ are both regular. Moreover,

$$(3\cdot6) \qquad\qquad \left| \frac{\partial x}{\partial y} \right| = 1 \Big/ \left| \frac{\partial y}{\partial x} \right|,$$

and therefore the inverse of a regular transformation is regular.

4. Continuous transformations.

Differential geometry is chiefly concerned with transformations of class $u > 0$. But for the sake of completeness we state three theorems about transformations of class 0.

If $x \rightarrow y$ is a non-singular continuous transformation of a region $[x]$ into a set of points $[y]$, then $[y]$ is also a region. This theorem is due to L. E. J. Brouwer, "Beweis der Invarianz des n-dimensionalen Gebietes," *Math. Annalen*, Vol. 71 (1912), pp. 305–13. See also H. Lebesgue, "Sur les correspondances entre les points de deux espaces," *Fundamenta Math.* Vol. 2 (1921), 256–85. A greatly simplified proof has been given by E. Sperner, "Neuer Beweis für die Invarianz der Dimensionszahl und des Gebietes," *Abhandlungen aus dem Math. Seminar der Hamburg. Universität*, Vol. 6 (1928), pp. 265–72. When $u > 0$ this theorem follows immediately from the implicit function theorem, as was shown in § 3.

If $x \rightarrow y$ is a non-singular continuous transformation of a region $[x]$ into a region $[y]$, the inverse transformation $y \rightarrow x$ is continuous (F. Hausdorff, *Mengenlehre*, Leipzig, 1914, Chap. IX, Theorem VIII).

Finally, it follows at once from the definition of continuous transformations that *the resultant of a continuous transformation which carries a region $[x]$ into a region $[y]$, followed by a continuous transformation of $[y]$ into $[z]$, is continuous.*

5. Pseudo-groups.

Instead of dealing, as in Chap. I, with transformations which carry the whole arithmetic space into itself, we now have to do with transformations operating on portions of the space. If $x \rightarrow y$ and $y' \rightarrow z$ are two such transformations which carry a set of points X into a set Y, and a set Y' into a set Z, respectively, the resultant $x \to z$ will exist if Y and Y' coincide, but only in this case. We are thus led to extend the notion of a transformation group to that of a pseudo-group. A set of transformations will be called a *pseudo-group* if it satisfies the conditions:

(i) *If the resultant of two transformations in the set exists it is also in the set.*

(ii) *The set contains the inverse of each transformation in the set.*

Clearly a transformation group in the arithmetic space is a special case of a pseudo-group.

For a given value of u (either 0, 1, ..., ∞ or ω) it follows from §§ 3 and 4 that the set of all regular transformations of class u is a pseudo-group. This will be called the *pseudo-group of class u*.

6. n-cells of class u.

The image (necessarily a region) of an n-dimensional box in a regular transformation of class u will be called an *arithmetic n-cell of class u*. Thus the arithmetic space is an n-cell, since it corresponds to the box $-1 < x^i < 1$ in the transformation

$$y^i = \tanh x^i.$$

It is obvious from the definition that all arithmetic n-cells of class u are equivalent under the pseudo-group of class u (i.e. that there exists a regular transformation which carries one of two given n-cells into the other); also that any set of points in a (1-1) correspondence of class u with an n-cell of class u is an n-cell of class u.

7. Simple manifolds of class u.

The pseudo-group of class u contains the group of all regular transformations which carry the arithmetic space into itself*. Let us denote this group by G_u. Any space satisfying the axioms G, given in Chap. II, § 9, with $G = G_u$, will be called a *simple manifold of class u*. Whenever G is a sub-group of G_u a space satisfying the axioms G is a simple manifold of class u plus additional structure (see Chap. II, § 9). Thus Euclidean spaces, affine spaces, and n-cells of class $u' > u$ are all simple manifolds of class u, whose additional structure is a smaller class of preferred coordinate systems than those described by the axioms G. Moreover, the same simple manifold can carry several spaces of a more complicated nature. Thus two affine spaces A_n and A_n' can have as

---

* This group contains the linear transformations and many others, such as
$$y^i = x^i + \tfrac{1}{2} \sin x^i.$$

their cartesian coordinate systems two distinct sub-sets of the preferred coordinate systems for the same simple manifold C_n. The straight lines for A_n and $A_n{}'$ respectively will be two systems of curves in C_n, each of which has all the affine properties of the arithmetic straight lines.

8. Oriented simple manifolds.

A transformation $x \rightarrow y$, of a region X into a region Y, will be called *direct* if the Jacobian $\left| \dfrac{\partial y}{\partial x} \right|$ is positive at all points of X, and *opposite* if $\left| \dfrac{\partial y}{\partial x} \right|$ is negative at all points* of X.

From the relations (3·5) and (3·6) it follows that the set of all direct transformations of class u is a pseudo-group.

This pseudo-group contains the group G_u^+, of all direct transformations of class u which carry the arithmetic space into itself. A space satisfying the axioms given in Chap. II, § 9, with $G = G_u^+$, will be called an *oriented simple manifold of class u*.

As in the case of affine spaces, two oriented manifolds are associated with any simple manifold C. The preferred coordinate systems for the oriented manifolds are obtained from an arbitrary preferred coordinate system for C by direct and opposite transformations respectively. The process of specifying either one of the oriented manifolds is called *orienting C*. Thus a simple manifold can be oriented in two, and only two, ways. The two oriented manifolds are equivalent under the group G_u.

9. Allowable coordinate systems for a simple manifold.

Though the structure of a space may be fully determined by a class of preferred coordinate systems, it is often convenient to use other coordinates which are well adapted to particular problems (e.g. polar and elliptic coordinates in Euclidean geometry). In this section we propose to define a class of "allowable" coordinate systems which is sufficiently wide for the ordinary purposes of differential geometry. Any allowable coordinate system is a (1-1) correspondence, $P \rightarrow x$, between a set of points, $[P]$, and a set of arithmetic points, $[x]$, in the arithmetic space of n dimensions. The set $[P]$ will be called the *domain*, and $[x]$ the *arithmetic domain*, of the coordinate system $P \rightarrow x$.

---

* A transformation may be neither direct nor opposite, for it may carry a region X_1 into Y_1 by a direct transformation and at the same time a region X_2 into Y_2 by an opposite transformation. This can only occur when X_1 and X_2 have no point in common.

Allowable coordinate systems of class u in a simple manifold of class u' ($u' \geqq u$) are those, and only those, which satisfy the following conditions:

1. *If* $[P]$ *is the image in a preferred coordinate system* K, *of an arithmetic region* $[x]$, *the correspondence* $P \to x$, *determined by* K, *is an allowable coordinate system.*

2. *If* $[x]$ *is an arithmetic region,* $P \to x$ *an allowable coordinate system, and* $x \to y$ *a regular transformation of class u, then the resultant transformation,* $P \to y$, *is an allowable coordinate system.*

If, in particular, the manifold is the arithmetic space with the transformations of G_u for preferred coordinate systems, the allowable coordinate systems are the transformations of the pseudo-group of class u. In general, the set of all transformations from allowable coordinate systems to allowable coordinate systems is the pseudo-group * of class u.

The image of a box in an allowable coordinate system of class u will be called an *n-cell of class u*. When the manifold is the arithmetic space, this reduces to the definition already given of an arithmetic n-cell. In the general case the image of a geometric n-cell in an allowable coordinate system is obviously an arithmetic n-cell, and *vice versa*. There is at least one allowable coordinate system in which a given n-cell †, C, is represented by the whole of the arithmetic space. With the set of all such allowable coordinate systems as preferred coordinate systems it may be verified from the conditions (1) and (2) that an n-cell satisfies the axioms for a simple manifold.

By associating an n-cell with the preferred coordinate systems which are obtained by direct transformations from a given allowable coordinate system in which the n-cell is represented, we obtain an *oriented n-cell*, which is an oriented manifold contained in the given manifold. Any orientation of the manifold obviously determines, and is determined by, an orientation of any n-cell contained in it.

The set of allowable coordinate systems of class u is independent of u', so long as $u' \geqq u$. To make this point clearer let us take an affine space A_n. To the cartesian coordinate systems for A_n let us adjoin all those obtained from a given cartesian coordinate system by transforma-

---

* This corrects an error in Q. F. Chap. II, § 2, where it is stated that the set of coordinate transformations is a group.

† We shall often omit the words "of class u" as applied to the various objects (e.g. transformations and n-cells) which we have defined. The class u and the dimensionality n will figure as parameters in the greater part of our discussion. Any discussion which involves derivatives will only apply to manifolds of a suitable class.

tions belonging to the group $G_{u'}$, for any $u' \geqq u$. We then have a set of preferred coordinate systems for a simple manifold of class u'. The same set of allowable coordinate systems is obtained by stating the above conditions, (1) and (2), in terms of either set of preferred coordinate systems.

The space A_n is an affine space in virtue of the fact that among the allowable coordinate systems is to be found a particular class of cartesian coordinate systems. Having admitted the allowable coordinate systems into our programme the central problem of affine geometry is to recover the cartesian coordinate systems, that is to characterize the latter as a sub-class of the former. We show how to do this in the next section. The methods we use only apply when $u \geqq 3$, and we suppose this to be the case.

10. The differential equations of affine geometry.

The domain of a cartesian coordinate system for an affine space, A_n, is the whole space, but the domain of an allowable coordinate system may be only a part of the space. Therefore we cannot, in general, talk about the transformation from allowable to cartesian coordinates, but must introduce the notion of a transformation *between* two coordinate systems.

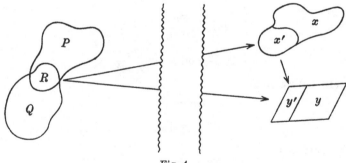

Fig. 4.

Let $P \to x$ and $Q \to y$ be two coordinate systems* and let $[R]$ be the intersection of $[P]$ and $[Q]$, that is to say the set of points common to $[P]$ and $[Q]$. If $[R]$ is not empty there are coordinate systems $R \to x'$ and $R \to y'$ in which each R corresponds to its image in $P \to x$ and in $Q \to y$, respectively. The transformation of coordinates $x' \to y'$ will be

---

* It is to be assumed that all coordinate systems referred to from now on are (1-1) correspondences $P \to x$, $[x]$ being a set of points in the arithmetic n-space.

called the transformation *between* $P \to x$ and $Q \to y$. If the set $[R]$ is empty the transformation between $P \to x$ and $Q \to y$ does not exist.

The problem stated at the end of the last section is to determine the transformation between a given allowable coordinate system and an arbitrary cartesian coordinate system.

The solution will have to depend on the fact that any two cartesian coordinate systems are related by a linear transformation. Suppose we represent coordinates in a given allowable system by x, in a particular cartesian system by u, and the transformation between them by

$$u^i = u^i(x).$$

The transformation to any cartesian coordinate system is given by

(10·1) $$y^i = a^i_j u^j(x) + a^i,$$

where the coefficients a are constants. The transformation from x to an arbitrary cartesian coordinate system will be characterized by the differential equations of which (10·1) are n independent solutions. To obtain these equations we write the general solution

$$y = a_j u^j(x) + a,$$

where a_i and a are arbitrary constants, and differentiate twice. We then have

(10·2)
$$
\begin{cases}
(a) & \dfrac{\partial y}{\partial x^j} = a_s u^s_j, \\[2ex]
(b) & \dfrac{\partial^2 y}{\partial x^j \partial x^k} = a_s u^s_{jk},
\end{cases}
$$

where $$u^s_j = \frac{\partial u^s}{\partial x^j} \text{ and } u^s_{jk} = \frac{\partial^2 u^s}{\partial x^j \partial x^k}.$$

From (10·2 a) we obtain

$$a_s = v^j_s \frac{\partial y}{\partial x^j},$$

where $$v_b u^a_j = \delta^a_b,$$

and from (10·2 b)

(10·3) $$\frac{\partial^2 y}{\partial x^j \partial x^k} - \frac{\partial y}{\partial x^i} \Gamma^i_{jk} = 0,$$

where

(10·4) $$\Gamma^i_{jk} = v^i_s u^s_{jk}.$$

Therefore the cartesian coordinate systems are given in each allowable coordinate system by a set of differential equations of the form (10·3). These will be called the *differential equations of affine geometry*.

If we make a transformation of coordinates $x \to \bar{x}$ the differential equations (10·3) go into

$$\frac{\partial^2 y}{\partial \overline{x}^j \, \partial \overline{x}^k} - \frac{\partial y}{\partial \overline{x}^i} \, \overline{\Gamma}^i_{jk} = 0,$$

where

(10·5) $$\overline{\Gamma}^i_{jk} = \left(\Gamma^a_{bc} \frac{\partial x^b}{\partial \overline{x}^j} \frac{\partial x^c}{\partial \overline{x}^k} + \frac{\partial^2 x^a}{\partial \overline{x}^j \, \partial \overline{x}^k} \right) \frac{\partial \overline{x}^i}{\partial x^a},$$

since any solution of (10·3) is to be a scalar (see *Q. F.* Chap. v, § 5). From the equations (10·4) it is evident that the functions Γ^a_{bc} are identically zero whenever the coordinate system $P \to x$ is cartesian. The formula (10·5) with $\Gamma^a_{bc} = 0$ then gives their value in any other coordinate system $P \to \overline{x}$. They are of class $u - 2$ in each allowable coordinate system.

Thus the structure of an affine space determines in every allowable coordinate system a set of functions Γ. From the relation (10·5) between the sets of functions Γ and $\overline{\Gamma}$, determined in different allowable coordinate systems, it follows that Γ are the components of an affine connection*. Any affine connection determined in this way by a flat affine space is said to be *flat*.

11. The differential equations of the straight lines.

The affine connection appears in various problems of flat affine geometry. For example, in cartesian coordinates the straight lines obviously satisfy the differential equations

$$\frac{d^2 y^i}{ds^2} = 0.$$

On transforming to arbitrary allowable coordinates, these differential equations become

$$\frac{d^2 x^i}{ds^2} + \Gamma^i_{jk} \frac{dx^j}{ds} \frac{dx^k}{ds} = 0,$$

where the functions Γ are those we have been discussing.

More generally, any flat k-space satisfies the differential equations

$$\frac{\partial^2 x^i}{\partial s^\lambda \, \partial s^\mu} + \Gamma^i_{jk} \frac{\partial x^j}{\partial s^\lambda} \frac{\partial x^k}{\partial s^\mu} = 0, \qquad (\lambda, \mu = 1, \dots, k).$$

When expressed as solutions of these equations, the k-spaces, and in particular the straight lines, are referred to affine parameterizations (Chap. ii, § 13).

---

* For a definition of an affine connection see *Q. F.* Chap. iii, § 10, or Chap. v, § 12, below. The affine connection itself, as distinguished from its components, is an abstract object whose existence we assume in order that something may have the components. It is entirely analogous from the logical point of view to any other abstract object of mathematics or physics, the number two, for example, or an electromagnetic field. See the footnote to *Q. F.* p. 16.

12. Integration of the differential equations of affine geometry.

The question now arises, what do we get by integrating the differential equations (10·3)? We assume the integrability conditions*

(12·1)
$$\begin{cases} \Gamma^i_{jk} = \Gamma^i_{kj}, \\ B^i_{jkl} = 0 \end{cases}$$

to be satisfied, where

(12·2)
$$B^i_{jkl} = \frac{\partial \Gamma^i_{jk}}{\partial x^l} - \frac{\partial \Gamma^i_{jl}}{\partial x^k} + \Gamma^s_{jk}\Gamma^i_{sl} - \Gamma^s_{jl}\Gamma^i_{sk}.$$

These are necessary and sufficient conditions in order that (10·3) admit a unique solution

(12·3)
$$u\,(x^1, \dots, x^n),$$

which satisfies given initial conditions

$$\begin{cases} u\,(x_0) = a, \\ (\partial u/\partial x^i)_{x_0} = a_i, \end{cases}$$

x_0 being any point near which (12·1) are satisfied. The solution (12·3) is a scalar function which is defined in some n-cell containing x_0. The general solution is

$$a_s u^s\,(x) + a,$$

where the a's are constants and $u^1\,(x), \dots, u^n\,(x)$ are n solutions whose Jacobian, $\left| \dfrac{\partial u}{\partial x} \right|$, is not zero at x_0.

The solutions $u^1, \dots, u^n$ are all defined in some n-cell containing x_0, and therefore an allowable coordinate system, $P \to y$, is obtained from $P \to x$ by any transformation of the form (10·1). Any such allowable coordinate system will be called a *locally cartesian coordinate system*. From § 10 it follows that locally cartesian coordinate systems are those, and only those, in which the components of the affine connection vanish.

The transformation between any two locally cartesian coordinate systems in which a given point is represented is linear, and any coordinate system obtained by a linear transformation from a locally cartesian coordinate system, is locally cartesian.

An affine connection will be called *locally flat* if, and only if, each point at which it is defined is represented in at least one locally cartesian coordinate system. This is the same as saying that the conditions (12·1) are satisfied wherever the affine connection is defined. A locally flat

---

* See *Q.F.* Chap. v, § 4. For a treatment which does not assume analyticity see F. Schur, *Math. Annalen*, Vol. 41 (1893), p. 509; or § 10 of the note referred to in Chap. vi, § 1, below.

affine connection is flat if, and only if, the region* over which it is defined is an n-cell, C, and if C is represented by the arithmetic space in
at least one locally cartesian coordinate system $P \rightarrow y$. In this case C
will be a flat affine space having $P \rightarrow y$ as a cartesian coordinate
system.

In general the region of definition need not be an n-cell. But it is a
space, according to Chap. II, § 1, whose structure is defined by the allowable coordinate systems in which it is represented, and by the locally
flat affine connection. Such a space will be called a *locally flat affine
space*, and its geometry *locally flat affine geometry*. Thus locally
flat affine geometry is the theory of a locally flat affine connection.

13. Three locally flat affine spaces†.

A locally flat affine structure is defined in the arithmetic space of two
dimensions by the system of curves

(13·1) $$a u(x, y) + b v(x, y) + c = 0,$$

where

(13·2) $$\begin{cases} u = e^x \cos y, \\ v = e^x \sin y, \end{cases}$$

a, b, and c being constants. Since $\dfrac{\partial (u, v)}{\partial (x, y)} \neq 0$, a locally flat affine connection is defined at each point by equations of the form (10·4). The
general solution of (10·3) is the left-hand side of (13·1), and the equations (13·2), with the condition $y_0 - \pi < y < y_0 + \pi$, define a transformation to a local cartesian coordinate system.

It is easy to see, however, that this is not a flat affine space. For
the locus (13·1) consists of a number of disconnected segments on account
of the periodicity in y of $u(x, y)$ and $v(x, y)$.

In our second example the space is a cylinder on which a locally flat
affine structure is defined by rolling it out on a plane. Each point, P,
on the cylinder corresponds to an infinite sequence, $[y_\alpha]$, of arithmetic
points, but a patch containing P is in a (1-1) correspondence with a
2-cell containing any one of the points y_α. Any such (1-1) correspondence
is to be taken as a local cartesian coordinate system. By imbedding
the cylinder in an affine 3-space, we can transform it analytically into
the ring given by

$$2 > x^2 + y^2 > 1,$$

---

* A region is the image in an allowable coordinate system of an arithmetic region.
Unless otherwise stated it is to be assumed that all affine connections and tensors to
which we refer are defined over regions.

† Cf. Chap. VII, § 3.

in an affine plane. The system of curves on the cylinder corresponding to the straight lines will be carried into a system of analytic curves in the ring, and will determine a locally flat affine connection at each point. But the general solution of the differential equations (10·3) will be a many-valued function when considered over the whole of the ring.

There are also locally flat affine spaces which are not equivalent to regions in a simple manifold. For example, a locally flat affine structure can be defined as follows on a space, R_2, which is topologically equivalent to an anchor ring. If (x, y) is any point in the arithmetic space of two dimensions the class of points given by

$$x + r, \quad y + s,$$

where r and s are any integers, positive, negative, or zero, is to be a single point in R_2. Any coordinate system $(x + r, y + s) \rightarrow (x, y)$ in which a 2-cell in R_2 is represented by a box

$$|x - x_0| < 1, \quad |y - y_0| < 1$$

is to be taken as a locally cartesian coordinate system.

The space R_2 belongs to the more general class of "regular" manifolds which we define in Chap. VI.

14. Geometric objects.

Anything which is unaltered by transformations of coordinates is called an invariant ($Q. F.$ Chap. II, § 2). Thus a point is an invariant and so is a curve or a system of curves. Also, strictly speaking, anything, such as a plant or an animal, which is unrelated to the space which we are talking about, is an invariant. For an invariant which is related to the space, i.e. a property of the space in the sense of Chap. II, § 1, we shall also use the term *geometric object* *.

A point is an example of a geometric object which determines a set of numbers in each allowable coordinate system in which it is represented †.

Other examples of geometric objects with components are affine connections and tensors of all kinds, and we now ask the reader to take for granted the elementary theory of these geometric objects as it is developed in $Q. F.$ Chap. II.

---

* This term was introduced as an alternative to "invariant" by J. A. Schouten and E. R. van Kampen, *Math. Annalen*, Vol. 103 (1930), p. 758.

† The totality of correspondences between a point, P, and its sets of coordinates is invariant. A transformation of coordinate systems, $x \rightarrow y$, does not alter the fact that the correspondences $P \rightarrow x$ and $P \rightarrow y$, as well as the others $P \rightarrow z$, $P \rightarrow t$, ... , all exist.

15. Regular point transformations.

A correspondence $P \to Q$ between a set of points $[P]$ and a set of points $[Q]$ will be called a *point transformation*, and will be said to carry $[P]$ into $[Q]$. A point transformation $P \to Q$ is a function,

$$Q = F(P),$$

whose argument is a point belonging to $[P]$ and whose values are the points $[Q]$. Unless otherwise stated it is to be assumed that any point transformation to which we refer is non-singular, meaning it is not only single-valued but also has a single-valued inverse.

Let $P \to Q$ be a point transformation between sets of points $[P]$ and $[Q]$ in the domain of the same coordinate system, K.

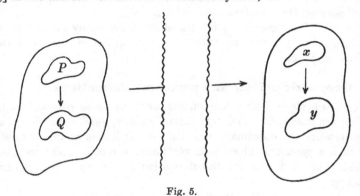

Fig. 5.

The transformation $P \to Q$ determines a transformation $x \to y$ in the arithmetic space, which carries the image, x, of each P into y, the image of the corresponding Q. Conversely a transformation $x \to y$ between sets of points, $[x]$ and $[y]$, in the arithmetic domain of the same coordinate system, determines a point transformation $P \to Q$.

A point transformation between sets of points both of which are represented in the same coordinate system, K, is said to be *represented* by the equations

$$y^i = y^i(x),$$

which define the corresponding transformation in the arithmetic space. The transformation $P \to Q$ may equally well be represented in any other coordinate system, $\overline{K}$, which contains both $[P]$ and $[Q]$ in its domain. For let the transformation of coordinates from K to $\overline{K}$ be defined by the equations

$$\bar{x}^i = f^i(x),$$

and let $\bar{x}$ and $\bar{y}$ be the images in $\bar{K}$ of P and Q respectively. Then the transformation $P \to Q$ is given in $\bar{K}$ by *

$$\bar{y}^i = f^i \{ y \left(f^{-1} \left(\bar{x} \right) \right) \}$$
$$= \bar{y}^i \left(\bar{x} \right).$$

A point transformation $P \to Q$, between regions $[P]$ and $[Q]$ in the domain of an allowable coordinate system K, will be described as a regular point transformation *of class* u' ($u' \leqq u$), if the transformation by which it is represented in K is regular and of class u'. The class of a point transformation is obviously an invariant.

By Chap. II, § 15, the group of automorphisms of a simple manifold consists of the regular point transformations which are given in preferred coordinates by the transformations of G_u.

The set of all regular point transformations is, by an analogous argument, simply isomorphic with the pseudo-group of class u in the arithmetic space.

16. Geometric objects and point transformations.

Geometric objects with components, such as affine connections or tensors, are classified by their transformation laws, which determine what happens when the coordinates are changed. It is also necessary to say how such a geometric object behaves under a regular point transformation. Let us denote the transformation law of such a geometric object by

$$(16 \cdot 1) \qquad \bar{\xi}^a \left(\bar{x} \right) = F^a \{ \xi^1 \left(x \right), \ldots, \xi^m \left(x \right), \bar{x} \left(x \right) \},$$

where $\xi^1, \ldots, \xi^m$ are its components in an allowable coordinate system, $P \to x$, and $\bar{\xi}^1, \ldots, \bar{\xi}^m$ are its components in an allowable coordinate system, $P \to \bar{x}$, given by

$$(16 \cdot 2) \qquad \bar{x}^i = \bar{x}^i \left(x \right).$$

Now let the equations $(16 \cdot 2)$ define a point transformation of a set of points $[x]$ into a set of points $[\bar{x}]$. Let ξ be any geometric object which is defined over $[x]$, and which has the transformation law given by $(16 \cdot 1)$. To ξ corresponds the geometric object $\bar{\xi}$ which is defined over $[\bar{x}]$, whose components are given by $(16 \cdot 1)$, and which has the same transformation law as ξ. The point transformation $x \to \bar{x}$ is said to carry ξ into $\bar{\xi}$. Under these circumstances ξ is said to be *equivalent* to $\bar{\xi}$.

---

* Denoting $x \to y$ by S, $x \to \bar{x}$ by T, and $\bar{x} \to \bar{y}$ by $\bar{S}$, we have
$$\bar{S} = TST^{-1},$$
where T^{-1} is applied first.

17. Geometric objects and their geometries.

Any geometric object in a simple manifold determines a particular structure and therefore a space in the sense of Chap. II, § 1. Such a space is given when we specify a particular geometric object, ξ, and its geometry will be called the *geometry of the object* ξ. Two objects ξ and $\bar{\xi}$ have the same geometry if, and only if, they are equivalent.

The geometric objects and the spaces which they determine are classified by means of the pseudo-group of regular point transformations, two objects belonging to the same class if, and only if, they are equivalent. It is this pseudo-group, rather than the group G_u, which is relevant, because a geometric object is not necessarily defined over the whole of a simple manifold (cf. the second example of a locally flat affine space in § 13). This classification of geometric objects is in the spirit of the Erlanger Programm, equivalence under the pseudo-group being an inevitable generalization of equivalence in the narrower sense.

The geometries referred to in the discussion of the Erlanger Programm and its proposed extensions at the end of Chap. II, § 16 are all theories of geometric objects. For example a Riemannian geometry is the theory of a symmetric tensor g_{ij} whose determinant does not vanish. A geometry defined by a geometric object falls within the categories of the Erlanger Programm if, and only if, the geometric object is characterized by its group of automorphisms, i.e. the group of point transformations which carry it into itself.

CHAPTER IV

CELLS AND SCALARS

1. Purpose of the chapter.

In this chapter we give some account of loci in a simple manifold of class u. This should serve as an adequate foundation for the theory of k-spaces in any particular manifold of n dimensions. The classical differential geometry, for instance, is the theory of curves and surfaces in a Euclidean metric 3-space. It is specially concerned with the theory of curves and surfaces "at a given point," that is with the metric structure of arbitrarily small 1-cells and 2-cells (see § 2 below), containing the given point.

All these theories rest upon the notion of a k-cell, to which the greater part of this chapter is devoted.

2. k-cells in n-space.

A *k-cell of class u* is any set of points given by

$$(2 \cdot 1) \qquad \begin{array}{ll} (a) & \{\, |\, y^\lambda\,| < 1, \qquad (\lambda = 1, \ldots, k) \\ (b) & \{\, y^\sigma = 0, \qquad (\sigma = k+1, \ldots, n) \end{array}$$

in an allowable coordinate system.

If we apply this definition to the case where our space is the arithmetic space of n dimensions it yields a definition of arithmetic k-cells. It follows from this definition that all arithmetic k-cells of class u are equivalent under the pseudo-group of class u (cf. Chap. III, § 6).

The more complicated the structure of a manifold, the greater will be the variety of the k-cells imbedded in it. Thus all k-cells in a manifold of class 0 are alike. In a manifold of class u there are k-cells of class 0, 1, $\ldots$, and u, while in a Euclidean space there are curved and flat k-cells.

The conditions $(2 \cdot 1)$ may also be written in the form

$$(2 \cdot 2) \qquad \begin{array}{ll} (a) & \{\, y^\lambda = s^\lambda, \\ (b) & \{\, y^\sigma = 0, \end{array}$$

where

$$(2 \cdot 3) \qquad |\, s^\lambda\,| < 1.$$

The relations $(2 \cdot 2)$ and $(2 \cdot 3)$ state that the k-cell is in (1-1) correspondence with the unit box in the arithmetic space of k dimensions. This correspondence is a coordinate system or parameterization, $P \to s$, for the k-cell. The k-cell with the parameterization is called a *parameterized k-cell*.

A parameterization $P \to s$ of a k-cell has been given as the resultant of an allowable coordinate system $P \to y$, followed by a transformation $y \to s$. Without altering the parameterization we can apply a transformation of coordinates, given by

$$(2\cdot4) \qquad x^i = f^i(y^1, \ldots, y^n).$$

The parameterization $P \to s$ is then given by (2·3) and

$$(2\cdot5) \qquad x^i = f^i(s^1, \ldots, s^k, 0, \ldots, 0).$$

Without altering the coordinate system we can also apply a transformation of parameter, given by

$$(2\cdot6) \qquad t^\lambda = t^\lambda(s^1, \ldots, s^k).$$

If $s \to t$ is a regular transformation of class u, it converts (2·5) into equations of the form

$$(2\cdot7) \qquad x^i = x^i(t^1, \ldots, t^k),$$

and the box (2·3) into an arithmetic k-cell of class u. Any parameterization obtained from the parameterization given by (2·2) by a regular transformation of class u will be called a *parameterization of class u*, or a *regular parameterization*. With the regular parameterizations in which it is represented by the arithmetic space of k dimensions as preferred coordinate systems, a k-cell is obviously a simple manifold of class u.

Now let us start with a set of n equations of the form (2·7) such that the functions on the right are of class u and independent in some arithmetic k-cell, T_k. By saying that the functions are independent we mean that the rank of the matrix

$$\left\| \frac{\partial x^i}{\partial t^\lambda} \right\|$$

is k. Without loss of generality we may assume that the notation is such that there is one set of values, $t_0^1, \ldots, t_0^k$, for which

$$(2\cdot8) \qquad \frac{\partial(x^1, \ldots, x^k)}{\partial(t^1, \ldots, t^k)} \neq 0.$$

From the continuity of the derivatives it follows that there is a box with centre t_0, in the arithmetic space of k dimensions, in which (2·8) holds good. This implies that there is a box in the arithmetic space of n dimensions in which the equations

$$(2\cdot9) \qquad x^i = x^i(t^1, \ldots, t^k) + \delta_\sigma^i(t^\sigma - t_0^\sigma), \qquad (\sigma = k+1, \ldots, n)$$

define a regular transformation $t \rightarrow x$. Hence the inverse of (2·9) defines a transformation to a coordinate system, t, in which an n-cell containing the point $x_0 = x(t_0)$ is imaged in a box with the arithmetic point $(t_0^1, \ldots, t_0^n)$ as centre. By an obvious transformation of this box we come back to an allowable coordinate system in which (2·7) reduces to (2·1). Hence, *if the functions on the right-hand side of* (2·7) *are independent at a point x_0, there is a k-cell, C_k, of class u containing x_0, whose points satisfy* (2·7). *Moreover, there is an n-cell containing x_0, such that all points in this n-cell which satisfy* (2·7) *lie in C_k.*

3. Implicit equations of a k-cell.

Let C_n be the n-cell which is the image of the box $-1 < y^i < 1$, in the coordinate system* y of the last section. An arbitrary coordinate system x, in which C_n is represented, is obtained from the coordinate system y by a transformation (2·4). In this coordinate system C_n is the image of an arithmetic n-cell of arbitrary type. The k-cell C_k consists of those points of C_n which satisfy (2·2b), and is described in the coordinates x by means of the $n - k$ equations,

$$(3·1) \qquad F^\sigma (x^1, \ldots, x^n) = 0, \qquad (\sigma = k + 1, \ldots, n),$$

the functions F being $n - k$ of those which represent the transformation

$$y^i = F^i (x^1, \ldots, x^n),$$

inverse to (2·4). The equations (3·1) are called the *implicit equations* of the k-cell, and (2·7) are called its *parametric equations*.

Each equation in the set (3·1) is an implicit equation to an $(n-1)$-cell. Hence C_k is represented as the intersection of $n - k$ $(n-1)$-cells.

Conversely to the theorem that any k-cell can be represented by equations of the form (3·1) we have: *Let the left-hand members of* (3·1) *be any $(n - k)$ functions of class u whose Jacobian matrix is of rank $n - k$ at a point x_0 which satisfies* (3·1). *There exists an n-cell, C_n, containing x_0, such that all the points in C_n which satisfy* (3·1) *lie on a k-cell, C_k, and each point of C_k satisfies* (3·1).

To prove this we may suppose without loss of generality that the notation is so assigned that the Jacobian

$$\frac{\partial (F^{k+1}, \ldots, F^n)}{\partial (x^{k+1}, \ldots, x^n)}$$

---

* In this and the next chapter we shall often drop into the somewhat freer terminology used in Q. F. Thus a coordinate system will mean an allowable coordinate system of class u. Of course anything we say which involves derivatives will apply only to spaces of an appropriate class. We may denote a point by the arithmetic point, x, to which it corresponds in a coordinate system, $P \rightarrow x$, and we may also use the single letter, x, to represent the coordinate system $P \rightarrow x$.

is not zero at x_0. In the arithmetic n-space there will therefore be a box having x_0 as its centre, which is carried by the transformation

$$(3\cdot2) \quad \begin{cases} y^\lambda = x^\lambda - x_0^\lambda, & (\lambda = 1, \ldots, k) \\ y^\sigma = F^\sigma(x) - F^\sigma(x_0), & (\sigma = k+1, \ldots, n) \end{cases}$$

into an arithmetic n-cell A_n, containing the origin. Let B_n be a box containing the origin and contained in A_n, and let B_k be the k-cell consisting of the points of B_n which satisfy the equations

$$y^\sigma = 0.$$

The inverse transformation to $(3\cdot2)$ carries B_n and B_k into an arithmetic n-cell and an arithmetic k-cell, respectively, which are the images in the coordinate system x, of the required n-cell, C_n, and k-cell, C_k.

4. Scalars.

Any point function, that is to say, any correspondence which associates a number $f(P)$ with each point of a set $[P]$, determines a function of n variables by means of the relation

$$F(x) = f(P)$$

in each coordinate system $P \to x$, in which the set $[P]$ is represented. The function $f(P)$ is called an *absolute scalar*, and the function $F(x)$ is called its *component* in the coordinate system $P \to x$. If the component of a scalar is of class $u' \le u$ in any one coordinate system it will be so in all of them. In this case $f(P)$ will be called a scalar of class u'. We shall assume that all the scalars referred to are of class u.

A scalar determines in any coordinate system the n derivatives,

$$\frac{\partial F}{\partial x^1}, \ldots, \frac{\partial F}{\partial x^n}.$$

These are the components of a geometric object called the *gradient* of the scalar. The gradient is a covariant vector, the components in a coordinate system $\bar{x}$ being connected with the components in a coordinate system x, by the equations

$$\frac{\partial \bar{F}}{\partial \bar{x}^i} = \frac{\partial F}{\partial x^j} \frac{\partial x^j}{\partial \bar{x}^i}.$$

Since these equations are linear homogeneous in the components, it is evident that if all the components of the gradient vanish at a point x_0 in one coordinate system, they all vanish at this point in all coordinate systems. In this case x_0 is called a *singular point* of the scalar, otherwise an *ordinary* point.

By the last section, if x_0 is an ordinary point of a scalar, and $F(x)$

the component of the scalar in any coordinate system x, an n-cell C_x can be found containing x_0, such that all points of C_x which satisfy the equation

$$F(x) = F(x_0)$$

constitute an $(n-1)$-cell. Moreover, a coordinate system, y, can be found in which the component of the scalar reduces to y^1. Thus the scalar can be thought of as determining in the neighbourhood of any point a layer of $(n-1)$-cells, like a family of parallel planes in a Euclidean space.

5. Sets of n − k scalars.

What has been said about one scalar generalizes at once to any set of $n-k$ scalars $(0 \leqq k \leqq n-1)$. The $(n-k)$-rowed determinants,

$$(5\cdot1) \qquad \frac{\partial(F^{k+1}, \ldots, F^n)}{\partial(x^{i_{k+1}}, \ldots, x^{i_n})} = e_{a_{k+1} \cdots a_n} \frac{\partial F^{a_{k+1}}}{\partial x^{i_{k+1}}} \cdots \frac{\partial F^{a_n}}{\partial x^{i_n}},$$

of the Jacobian matrix are the components of a geometrical object called a *generalized gradient*. It is obviously a covariant tensor, and if all the components of the generalized gradient vanish at a particular point in a given coordinate system, they will vanish at this point in all coordinate systems. Such a point is called a *singular point* of the set of scalars. Any other point at which the scalars are all defined is called an *ordinary point*. From the continuity of the derivatives it follows that the set of all ordinary points is a region.

Just as in the case of a single scalar, if x_0 is an ordinary point, and $F^\sigma(x)$, $(\sigma = k+1, \ldots, n)$, are the components of the $n-k$ scalars in any coordinate system in which x_0 is represented, there exists an n-cell C_x, such that all points of C_x which satisfy the equations

$$(5\cdot2) \qquad F^\sigma(x) = F^\sigma(x_0)$$

constitute a k-cell. Moreover, a coordinate system can be found in which the components of the scalars reduce to $y^{k+1}, \ldots, y^n$ respectively.

6. Sets of scalars and oriented k-cells.

Any set of n scalars $f^1(P), \ldots, f^n(P)$ in a definite order determines a coordinate system, $P \to y$, given by

$$(6\cdot1) \qquad y^i = f^i(P).$$

The transformation $P \to y$ need not be (1-1), but is a coordinate system in the general sense of Chap. II, § 8. If the scalars are of class u, and if Q is an ordinary point, the equations $(6\cdot1)$ define an allowable coordinate system, y, in which some n-cell, C_Q, containing Q is represented. If x is any allowable coordinate system in which Q is represented, the transformation between x and y is given by

$$y^i = F^i(x),$$

where $F^i(x)$ is the component of $f^i(P)$ in x.

The n-cell C_Q, associated with the totality of parameterizations obtained from (6·1) by direct transformations of coordinates, is an oriented simple manifold according to Chap. III, § 8. Hence an orientation of C_Q is determined by the n scalars $f^1(P), \ldots, f^n(P)$ in a definite order. This orientation is changed into the opposite one if any one of these scalars is replaced by its negative. It is unchanged by even, but altered by odd permutations of the n scalars. This is the gist of the statement sometimes made that an orientation is determined by giving an order to the coordinates.

Any point, P_0, in C_Q is the intersection of the k-cell, C_k, given by

$$(6·2) \qquad \begin{cases} (a) & f^\lambda(P) = s^\lambda, & (\lambda = 1, \ldots, k), \\ (b) & f^\sigma(P) = f^\sigma(P_0), & (\sigma = k+1, \ldots, n), \end{cases}$$

and the $(n-k)$-cell, C_{n-k}, given by

$$(6·3) \qquad \begin{cases} (a) & f^\lambda(P) = f^\lambda(P_0), \\ (b) & f^\sigma(P) = s^\sigma. \end{cases}$$

These two cells may be described as *dual* to each other. The equations (6·2 a) define a parameterization for C_k, and it follows by the argument used above that C_k may be oriented by assigning a sense to the ordered set of scalars $f^1(P), \ldots, f^k(P)$. An orientation thus defined will be called an *interior* orientation of C_k. Similarly an interior orientation is defined for C_{n-k} by assigning a sense to the ordered set of scalars $f^{k+1}(P), \ldots, f^n(P)$. The orientation of two of the cells, C_k, C_{n-k} and C_Q, determines an orientation of the third by means of the relation

$$(6·4) \qquad \text{sign } C_k \cdot \text{sign } C_{n-k} = \text{sign } C_Q.$$

Let $\bar{f}^{k+1}(P), \ldots, \bar{f}^n(P)$ be any other ordered set of scalars which, equated to suitable constants, give a set of implicit equations to C_k. Let

$$F^\sigma(x) = f^\sigma(P),$$
$$\bar{F}^\sigma(x) = \bar{f}^\sigma(P),$$

in any coordinate system whose domain contains C_k. At any point on C_k

$$\frac{\partial F^{k+1}}{\partial x^i}, \ldots, \frac{\partial F^n}{\partial x^i} \text{ and } \frac{\partial \bar{F}^{k+1}}{\partial x^i}, \ldots, \frac{\partial \bar{F}^n}{\partial x^i}$$

are two complete sets of solutions to the algebraic equations in u,

$$\frac{\partial x^i}{\partial t^\lambda} u_i = 0, \qquad (\lambda = 1, \ldots, k)$$

where

$$x^i = x^i (t^1, \ldots, t^k)$$

are parametric equations for C_k. From Chap. I, §§ 3 and 5 it follows that there are numbers a_ρ^σ such that

$$\frac{\partial \overline{F}^\sigma}{\partial x^i} = a_\rho^\sigma \frac{\partial F^\rho}{\partial x^i}, \qquad (\rho, \sigma = k+1, \ldots, n)$$

at any point on C_k. The determinant $|a_\rho^\sigma|$ is a scalar which is defined over C_k, and does not vanish. If it is positive the sets of scalars f and $\overline{f}$ will be said to be positively related with respect to C_k, and negatively related otherwise. The relation of being positively related is obviously transitive, and if f is positively related to $\overline{f}$, and $\overline{f}$ negatively related to a set of scalars $\tilde{f}$, then f is negatively related to $\tilde{f}$. Therefore the ordered sets of scalars which enter in implicit equations for C_k fall into two classes, members of the same class being positively related to each other. These classes may be called *exterior sense-classes*, or *exterior orientations* of C_k.

As in the case of interior orientation an exterior orientation of C_k is determined by assigning a positive sense to the ordered set of scalars $f^{k+1}(P), \ldots, f^n(P)$, where (6·2b) are implicit equations for C_k. Therefore an exterior orientation determines, and is determined by, an interior orientation of a given $(n-k)$-cell which is dual to C_k. If $k = n-1$, for example, an exterior orientation is the association of the positive sign of $f^n(P)$ with one side of C_k, and of the negative sign with the other. Any 1-cell which meets C_k just once may be oriented by describing points on the negative side of C_k as before points on the positive side.

From the preceding paragragh it follows that the two interior, and the two exterior orientations of C_k, and the two orientations of C_Q, are related by an equation analogous to (6·4). For example, an oriented 3-cell may be represented by a right-handed screw, and a screw may either be pushed in to make it rotate, rotated to make it penetrate, or it may be determined as the resultant of a rotation and a translation along the axis of the rotation.

7. k-spaces in the large.

The locus which satisfies $n-k$ scalar equations of the form

$$(7\cdot1) \qquad F^{k+1}(x) = C^{k+1}, \ldots, F^n(x) = C^n,$$

C^{k+1}, ..., C^n being constants, has now been characterized as being made up of *k*-cells, provided we exclude all singular points. Moreover, we can say something about how these *k*-cells are pieced together to make the whole locus, namely that *if two of these k-cells have a point P in common, there is a k-cell containing P and contained in each of the given k-cells.*

Proof: Each of the given *k*-cells, according to the way they are derived in § 3, is associated with an *n*-cell and contains all the points common to the locus (7·1) and this *n*-cell. Let C_n be an *n*-cell containing *P* and contained in each of these *n*-cells. By § 3 we can find an *n*-cell, C_n', contained in C_n, containing *P*, and such that the set of points in C_n' which satisfy (7·1) constitute a *k*-cell. This last *k*-cell is the one whose existence we were to prove.

We can also prove that *if P and Q are any two ordinary points of the locus (7·1) there are two k-cells, one containing P, the other containing Q, both consisting entirely of points of the locus (7·1), and having no point in common.* To see this it is only necessary to take two *n*-cells containing *P* and *Q* respectively and having no point in common. By § 3, each of these *n*-cells contains a *k*-cell whose points satisfy (7·1), one *k*-cell containing *P* and the other containing *Q*.

The two theorems of this section together with § 2 are the essential points in proving that the set of ordinary points of the locus (7·1) is what, in Chap. VI below, is called a regular manifold.

8. Local properties. Infinitesimal geometry.

A property will be said to be *local* to a point *P* if there is an *n*-cell C_P, containing *P*, such that the property is common to all regions containing *P* and contained in C_P. The system of local properties at *P* will be called the *local structure* at *P*, and the theory of the local structure the *infinitesimal geometry** at *P*. Thus the infinitesimal geometry at *P* will contain theorems about the relations between *P* and points near *P*, but will involve no statement about a specified point other than *P*.

For instance *P* might lie at one end of the longest axis of an ellipsoid. The infinitesimal geometry of the surface at *P* contains the theorem that the Gaussian curvature at *P* is a minimum. But this is not a theorem in the infinitesimal geometry at an arbitrary point of the surface.

---

* This term is used in this sense by many writers. Others use "differential geometry" where we use "infinitesimal geometry." We use differential geometry in a wider sense, which is defined in Chap. VI, § 9.

In infinitesimal geometry the equivalence problem of Chap. III, §§16 and 17, is replaced by the problem of local equivalence. A geometrical object ξ will be called *locally equivalent* at P to a geometrical object $\bar{\xi}$ at $\bar{P}$ if, and only if, ξ and $\bar{\xi}$ are equivalent in n-cells containing P and $\bar{P}$ respectively. All locally flat affine connections, for example, are locally equivalent, but not equivalent.

It is the problem of characterizing locally equivalent geometric objects which is considered in *Q. F.* Chap. V. The solution is given for affine connections and for quadratic differential forms.

9. Equivalence of scalars.

The local equivalence problem for scalars at ordinary points is solved in §4 above. For suppose we have given two scalars, and let P be an ordinary point for the first and Q for the second. It is clear that the two scalars cannot be equivalent under a transformation which carries P into Q unless the first has the same value at P as the second at Q. Suppose now that this condition is satisfied. Then by §3 there exists a coordinate system, y, in which P is represented, and in which the first scalar has the component y^1. Likewise there exists a coordinate system, z, in which Q is represented, and in which the second scalar has the component z^1. For a sufficiently small n-cell containing P, a point transformation is defined by requiring each point, y, of this cell to go to that point whose z coordinates are given by

$$z^i = y^i.$$

This point transformation carries the first scalar in a cell containing P into the second scalar in a cell containing Q.

In like manner, by §5, if P and Q are ordinary points for two sets of $n - k$ scalars ($0 \leqq k \leqq n - 1$) there is a point transformation carrying P, and a cell enclosing it, into Q and a cell enclosing it, and transforming the first set of scalars into the second set if, and only if, the two sets of scalars take on the same values at P and Q.

This discussion illustrates the distinction between local equivalence and equivalence. For a scalar which, equated to a suitable constant, gives a closed surface (e.g. $x^2 + y^2 + z^2$) is not equivalent to a scalar which only gives open surfaces (e.g. $x + y + z$).

CHAPTER V

TANGENT SPACES

1. Differentials of a function.

A k-cell of class 1 in the arithmetic n-space is distinguished among the k-cells of class 0 by the existence of a tangent flat k-space at each point. Similarly any simple manifold of class 1 has a "tangent space" at each point, which may be defined without imbedding the manifold in a space of higher dimensionality. In this chapter we give some account of the tangent spaces, as a preliminary to which we develop the theory of differentials more fully than is done in $Q.F.$

A regular function of n variables, $F(x^1, \ldots, x^n)$, determines a function of $2n$ variables

$$dF = \frac{\partial F}{\partial x^i}\, dx^i,$$

which is of class $u - 1$ in x and linear homogeneous in dx. The variables $dx^1, \ldots, dx^n$ are called the *differentials* of the variables $x^1, \ldots, x^n$, respectively, and dF is called the *differential* of $F(x)$. If in particular $F(x) = x^i$ the function dF reduces to dx^i.

The function dF is not necessarily a "small quantity" though it tends to zero with dx, like any other linear homogeneous function*. Moreover, dF represents, to a first order approximation, the change in $F(x)$ due to a change from x to $x + dx$. For by an elementary theorem of differential calculus

$$F(x + dx) - F(x) = \frac{\partial F}{\partial x^i}\, dx^i + \epsilon\,(x,\, dx),$$

where

$$\underset{t \to 0}{\mathrm{Lim}}\; \frac{\epsilon\,(x,\, dx)}{t} = 0,$$

if $dx^i = p^i t$.

If $u \geqq 2$ there is also a function of $4n$ variables, called the *second differential* of $F(x)$,

$$\delta dF = \frac{\partial F}{\partial x^i}\, \delta\, dx^i + \frac{\partial^2 F}{\partial x^i \partial x^j}\, dx^i\, \delta x^j.$$

This is of class $u - 2$ in x and linear in each of the sets of n variables dx, δx, and δdx. It is the first differential of the function dF of $2n$

---

* Indeed dF is defined for all values of dx, though it may only be defined for values of x in a restricted region.

variables x and dx, the first differentials of these variables being now represented by δx and δdx respectively. By the same theorem of calculus as before, it represents to a first order approximation the change in dF due to a change from x and dx, to $x + \delta x$ and $dx + \delta dx$, respectively. When $F(x) = x^i$, the function δdF reduces to δdx^i.

If $u \geqq 3$ there is a function of $8n$ variables, the first differential of δdF, called the third differential of F,

$$\Delta \delta dF = \frac{\partial F}{\partial x^i} \Delta \delta dx^i + \frac{\partial^2 F}{\partial x^i \partial x^j} (\Delta dx^i \, \delta x^j + dx^i \, \Delta \delta x^j + \delta dx^i \, \Delta x^j)$$

$$+ \frac{\partial^3 F}{\partial x^i \partial x^j \partial x^k} dx^i \, \delta x^j \, \Delta x^k.$$

In like manner we can define fourth differentials, and so on.

2. Transformations of differentials.

The various sets of successive differentials of the variables x are simply additional sets of n variables each. Each set of values of each set of variables is a point in the arithmetic space of n dimensions. A regular transformation

$$(2\cdot1) \qquad\qquad y^i = y^i(x)$$

determines a transformation

$$(2\cdot2) \qquad\qquad dy^i = \frac{\partial y^i}{\partial x^j} dx^j.$$

If the point x be fixed, and dx be variable, this is a linear transformation of the arithmetic space of n dimensions into itself.

Equally well, $(x^1, \ldots, x^n, dx^1, \ldots, dx^n)$ is a point in the arithmetic space of $2n$ dimensions. The two sets of equations $(2\cdot1)$ and $(2\cdot2)$ together define a transformation in the arithmetic space of $2n$ dimensions. The set of all these transformations in the arithmetic space of $2n$ dimensions, determined by the transformations $(2\cdot1)$ of any pseudo-group, G^0, is obviously a pseudo-group, G^1. This new pseudo-group is called the first extension* of G^0. The higher extensions may be defined recursively. The kth extension of G^0 is a pseudo-group, G^k, in the arithmetic space of $2^k \cdot n$ dimensions, and G^{k+1} is the first extension of G^k, for $k = 0, 1, \ldots, u - 1$.

3. Differentials at a point.

We return now to the geometry of a simple manifold. Let

$$dx = (dx^1, \ldots, dx^n)$$

* S. Lie, *Theorie der Transformationsgruppen*, erster Abschnitt, Leipzig, 1888, p. 525.

be any arithmetic point associated with a point P and a coordinate system, x, in which P is represented. The equations

$$(3\cdot1) \qquad\qquad dy^i = \left(\frac{\partial y^i}{\partial x^j}\right)_P dx^j$$

define a transformation $dx \to dy$ of dx into an arithmetic point dy, associated with P and with any coordinate system y, in which P is represented. The geometric object determined by this association of an ordered set of numbers $dy^1, \dots, dy^n$ with P and each coordinate system y, is called a *contravariant* vector at P, or a *differential* at P (cf. Q.F. Chap. II, § 5). The numbers $dy^1, \dots, dy^n$ are called its *components* in y, and the equations $(3\cdot1)$ are said to define its *transformation law*. The transformation $(3\cdot1)$ is single-valued, and therefore no two distinct differentials have the same components in any coordinate system.

4. Tangent spaces.

The totality of differentials at any point P is called the *tangent space* of differentials at P. Thus a simple manifold has a tangent space at each point, and will therefore be referred to as the *underlying manifold*. Any coordinate system, x, for the underlying manifold determines a unique coordinate system, dx, for the tangent space at each point which is represented in x. Moreover, the coordinate system dx is a (1-1) correspondence between the tangent space and the arithmetic space of n dimensions. For any arithmetic point dx may be taken as the set of components in x of a differential at a given point P, and on the other hand the components dx of any differential are the components of an arithmetic point.

With the coordinate systems, dx, dy, $\dots$, corresponding to the various coordinate systems, x, y, $\dots$, as preferred coordinate systems, the tangent space at P satisfies the axioms given in Chap. II, § 9, for a centred affine space of n dimensions. For the transformation $dx \to dy$, between any two of these coordinate systems is given by $(3\cdot1)$, and is linear and homogeneous. Moreover, there are transformations of coordinates, $x \to y$, which determine a given linear homogeneous transformation of coordinates, $dx \to dy$, for the tangent space at P.

5. Oriented tangent spaces.

Consider any 1-cell of class 1. If it is oriented, the tangent at each point is oriented (we may think of each tangent as marked with an arrow). Conversely, if the tangent at a single point is oriented, an orientation is determined for the 1-cell, and hence for the tangent at every other point. These observations generalize as follows to the theory of simple manifolds in general.

A necessary and sufficient condition that a transformation of coordinates $x \to y$ be direct, is that the corresponding transformations of coordinates $dx \to dy$, for the tangent spaces at points in the domain of x, be direct. For the latter are defined by (2·2), and are direct if, and only if, the Jacobian is positive. It follows that each tangent space to an oriented manifold is an oriented centred affine space. Conversely, an orientation of a single tangent space to a simple manifold determines a class of preferred coordinate systems which are related to each other by direct transformations. That is to say an orientation for the manifold is determined, and hence an orientation for all the other tangent spaces.

6. Approximate flatness near a given point.

Let x' be any point near a given point x, and let $\epsilon^i(x')$ be functions of class $u-1$, which are small quantities of order $1+\delta$, $(\delta > 0)$ as x' tends to x. The equations

$$(6·1) \qquad dx^i = x'^i - x^i + \epsilon^i(x')$$

define a transformation $x' \to dx$ from some n-cell containing x, in the underlying manifold, to an n-cell in the tangent space. Let y be any other coordinate system obtained from x by a transformation $x \to y$, and let y'^i be the coordinates of x' in y. From (3·1) and the theorem of differential calculus referred to in §1, it follows that the transformation given in y by

$$(6·2) \qquad dy'^i = y'^i - y^i + \eta^i(y'),$$

where $\eta^i(y')$ aré small quantities of order $1+\delta$, agrees with that given by (6·1) as far as first order quantities are concerned. Therefore the class of transformations given by equations of the form (6·1) from an n-cell containing a point, x, to the tangent space at x is an invariant*. The point x in the underlying manifold is carried by each of these transformations into the origin or null-vector in the tangent space, and is called the *point of contact* with the tangent space. The centre of any tangent space may be said to coincide with the point of contact in the underlying manifold.

The equations (6·1) may also be taken to define a transformation of coordinates in the underlying manifold. Each coordinate system dx so obtained is called a *local coordinate system* at the point x. The local coordinate system dy at the same point for any other coordinate system

* It is this invariant class of transformations which justifies the term tangent space.

K_y is obtained from dx by a transformation which agrees with (3·1) as far as first order terms are concerned. This approximate flatness, or "smoothness," distinguishes an n-cell of class 1 from a general n-cell of class 0.

If the underlying manifold happens to be a flat affine space, A_n, the equations

$$dy^i = y^i - y_0^i,$$

where y are cartesian coordinates, determine an invariant transformation of the centred affine space, A_n^0, obtained by taking y_0 as the centre of A_n, into the tangent space, at y_0. For y and dy are cartesian coordinate systems for A_n, and for the tangent space, respectively. Therefore the tangent space at y_0 may be said to coincide with A_n^0, the centre of the tangent space coinciding with y_0.

7. Differentials and k-cells.

A k-cell, C_k, may be represented in a coordinate system, x, by parametric equations of the form

(7·1) $x^i = x^i (t^1, \dots, t^k).$

The values of the differentials of the functions $x^i (t)$ at any point $x_0 = x (t_0)$, in C_k, are the components of a differential in the tangent space at x_0. The totality of such differentials will be the parameterized linear k-space E_k (Chap. I, § 3), given by

(7·2) $dx^i = \left(\dfrac{\partial x^i}{\partial t^\lambda}\right)_0 dt^\lambda, \qquad (\lambda = 1, \dots, k).$

The k-cell, C_k, determines E_k, and the parameterization t for C_k determines the parameterization, dt, for E_k. The k-cell, C_k, associated with its regular parameterizations, is a simple manifold of k dimensions (Chap. IV, § 2), and the linear k-space E_k may be identified with the tangent space to C_k at t_0.

Conversely, let a linear k-space, $E_k(x)$, be defined by

(7·3) $dx^i = \xi_\lambda^i (x) dt^\lambda$

at each point, x, in some region X. Does there exist a family of k-cells, one, and only one, of which passes through each point x, and has $E_k (x)$ for its tangent space?

The answer is yes, if, and only if, the differential equations

$$\xi_\lambda^i \frac{\partial \phi}{\partial x^i} = 0$$

are completely integrable*, for then they will have $n - k$ independent

* See E. Goursat, *Cours d'analyse*, Vol. 2, Chap. XXII, § 450.

solutions $\phi^{k+1}(x), \ldots, \phi^{n}(x)$, and the required k-cells will be given by the implicit equations

$$\phi^{\sigma}(x) = \phi^{\sigma}(x_0).$$

Does there exist a set of parameterized k-cells which have the parameterized k-spaces given by (7.3) as tangent spaces, and whose parameterizations determine these parameterizations of the k-spaces?

The answer is yes, if, and only if, the differential equations

$$\frac{\partial x^i}{\partial t^\lambda} = \xi^i_\lambda(x)$$

are completely integrable *. When $k = 1$ the answer is always yes.

8. Geometry of the tangent spaces.

The simplest geometric interpretation of tensors is in connection with the geometry of the tangent spaces. By the definition of the tangent spaces, a contravariant vector field $V(x)$ determines a point

$$dx^i = V^i(x)$$

in the tangent space at each point at which it is defined. Any covariant vector field, defined over a set of points $[x]$, determines a linear differential form

$$A_i\, dx^i,$$

which, equated to a constant, gives the equation to a hyperplane in the tangent space at each x. A tensor of the second order determines a quadratic differential form

$$(8.1) \qquad g_{ij}\, dx^i\, dx^j,$$

which, equated to a constant, gives the equation to a quadric whose centre is at the origin. Each contravariant vector, X^i, determines a unique covariant vector given by

$$(8.2) \qquad X_i = g_{ij} X^j.$$

The hyperplane

$$(8.3) \qquad X_i\, dx^i = 1$$

is the polar of X^i with respect to the quadric

$$(8.4) \qquad g_{ij}\, dx^i\, dx^j = 1.$$

If the form is non-degenerate (i.e. if the determinant $|g_{ij}|$ is not zero) any covariant vector, X_i, determines a unique contravariant vector, given by

$$X^i = g^{ij} X_j,$$

* See the footnote to p. 44.

which is the pole of the hyperplane (8·3) with respect to the quadric
(8·4). This is the geometric interpretation of the formal operations of
lowering and *raising* indices.

The quadratic form (8·1) may be taken to define the length of the
vector dx. If it is positive definite (i.e. never negative) it defines a
centred Euclidean metric in each tangent space. For there is a class of
coordinate systems y in which the components of g_{ij} at a given point
P_0 are δ_{ij}. The corresponding coordinate systems, dy, may be taken
as rectangular cartesian coordinate systems in the tangent space at P_0,
and the length of a vector dy is given by

$$(dy^1)^2 + \dots + (dy^n)^2.$$

A mixed tensor of the second degree, A_j^i, determines a collineation,

$$\delta x^i = A_j^i \, dx^j,$$

in the tangent space at each point at which it is defined. It also defines
an infinitesimal transformation

$$\delta x^i = dx^i + \epsilon A_j^i \, dx^j.$$

If

$$g_{is} A_j^s + g_{sj} A_i^s = 0,$$

this is an infinitesimal Euclidean displacement.

These examples must suffice. For a systematic account of the algebraic theory of tensors see J. A. Schouten, *der Ricci-Kalkül*, Berlin,
1922, Chap. I.

9. Other coordinates in the tangent spaces.

It is sometimes desirable to use other coordinates than dx^i, dy^i, etc.,
in the tangent spaces. If, for example, v_i^a ($a = 1, \dots, n$) are n covariant
vectors, such that the determinant

$$|v_i^a|$$

does not vanish, a coordinate system is defined by the transformation
$dx \to \omega$, given by

(9·1) $$\omega^a = v_i^a \, dx^i.$$

This is a cartesian coordinate system, but unlike the coordinate system
dx, which undergoes the transformation (2·2) when the underlying
coordinates undergo (2·1), it is unaltered by transformations of the
underlying coordinates. Thus ω^a are scalars and if $F^a(x)$ are any n
scalars,

$$\omega^a = F^a(x)$$

designates a single point in the tangent space at each point where the scalars are defined.

In a Riemannian space v_i^a may be mutually orthogonal unit vectors. The coordinate systems ω, given by (9·1), will then be rectangular cartesian. This is the basis of a method originated by Ricci and Levi-Civita and subsequently used by many authors. See, for example, L. P. Eisenhart, *Riemannian Geometry*, Chap. III. The more general coordinate systems given by (9·1) are now used in various ways, especially by Cartan and Schouten, in connection with the generalizations of Levi-Civita's parallel displacement. They are systematically described by R. Lagrange in his Mémorial volume, *Calcul différentiel absolu*, Paris, 1926.

10. Tangent and osculating Riemannian spaces.

For the sake of clear distinctions we mention here another way in which the term tangent space can be used. Two tensors, g_{ij} and G_{ij}, in the same regular n-cell determine two Riemannian spaces which are identical as sets of points but which have different structures. Suppose that

$$(10·1) \qquad\qquad g_{ij}(x_0) = G_{ij}(x_0), $$

at a given point x_0, in a given coordinate system. Since g and G are tensors this equality will hold in all coordinate systems in which x_0 is represented. The two Riemannian spaces are said to be *tangent* to each other at x_0. They both determine the same geometry in the tangent space at x_0.

The Riemannian spaces determined by g and G are said to have *contact of order k*, if in addition to the relation (10·1) all derivatives of g_{ij} of order less than or equal to $k-1$ are equal to the corresponding derivatives of G_{ij} at the point x_0. In case $k=2$ the two Riemannian spaces are said to *osculate*. In this case the affine connections (Christoffel symbols) determined by the two spaces (Q. F. Chap. III, § 9) have the same components at x_0.

Assuming the quadratic form (8·1) to be positive, and $k=1$ or 2, one of the Riemannian spaces can be taken to be Euclidean, and the relation of tangency or osculation used to carry over the results of Euclidean to Riemannian geometry. On this subject the reader is referred to the book by E. Cartan, *Leçons sur la géométrie des espaces de Riemann*, Paris, 1928. This method gets its fullest results when the Euclidean space of closest contact is used, namely that one in which the cartesian coordinates at x_0 are the same as the Riemannian normal coordinates (*Q. F.* Chap. VI, § 16) for the Riemannian space.

11. Second differentials.

We have as yet said nothing about relations between tangent spaces at different points of the underlying space. We come to such relations when we consider second differentials. With a point x, an ordered pair of first differentials dx, δx, and a coordinate system x, let us associate an arithmetic point $\delta dx = (\delta dx^1, \ldots, \delta dx^n)$. The equations

$$(11\cdot1) \qquad \delta dy^i = \frac{\partial y^i}{\partial x^j}\, \delta dx^j + \frac{\partial^2 y^i}{\partial x^j \partial x^k}\, dx^j\, \delta x^k$$

define a transformation $\delta dx \to \delta dy$ into an arithmetic point δdy associated with x, the differentials dx, δx, and any coordinate system y in which x is represented. The geometric object determined by this association of an arithmetic point, δdx, with x, dx, δx, and each coordinate system, is called a *second differential associated with x, dx, and δx*.

Notice that the difference $\xi_2 - \xi_1$ between any two second differentials ξ_1 and ξ_2, associated with x and the differentials dx, δx, is a first differential. For any transformation of coordinates $x \to y$ carries ξ_λ into η_λ ($\lambda = 1, 2$), given by $(11\cdot1)$, where ξ_λ and η_λ are written for δdx and δdy respectively. Therefore

$$\eta_2^i - \eta_1^i = \frac{\partial y^i}{\partial x^j}\, (\xi_2^j - \xi_1^j).$$

The same argument shows that

$$\delta dx - d\, \delta x$$

is a first differential. Hence it is an invariant condition to equate this difference to an arbitrary contravariant vector. In particular, the equation

$$\delta dx - d\, \delta x = 0,$$

which is assumed in many problems, is an invariant condition.

12. Affine connections.

Let a second differential be associated with each point, x, in some region, and each ordered pair of first differentials at x. In any coordinate system, x, let the components of the second differential associated with x, dx and δx be given by

$$(12\cdot1) \qquad \delta dx^i = \gamma^i\,(x, dx, \delta x).$$

The functions γ^i are the components of a geometric object which may be called a *field* of second differentials (in accordance with the usage in

Q. F. Chap. II) because it determines one second differential for each x, dx, and δx.

Let γ^i be linear homogeneous in δx^i. The equations (12·1) may then be written

$$(12\cdot2) \qquad \delta\,dx^i = \gamma^i_k\,(x,\,dx)\,\delta x^k.$$

In any other coordinate system $\bar{x}$ the components of this second differential are given by

$$\delta\,d\bar{x}^i = \frac{\partial\bar{x}^i}{\partial x^j}\,\gamma^i_k\,\delta x^k + \frac{\partial^2\bar{x}^i}{\partial x^j\,\partial x^k}\,dx^j\,\delta x^k$$

$$= \bar{\gamma}^i_k\,(\bar{x},\,d\bar{x})\,\delta\bar{x}^k,$$

where

$$(12\cdot3) \qquad \bar{\gamma}^i_k = \left(\gamma^a_b\,\frac{\partial x^b}{\partial\bar{x}^k} - \frac{\partial^2 x^a}{\partial\bar{x}^j\,\partial\bar{x}^k}\,d\bar{x}^j\right)\frac{\partial\bar{x}^i}{\partial x^a},$$

as follows from the identities

$$\frac{\partial^2 x^a}{\partial\bar{x}^j\,\partial\bar{x}^k}\,\frac{\partial\bar{x}^j}{\partial x^b}\,\frac{\partial\bar{x}^k}{\partial x^c} + \frac{\partial x^a}{\partial\bar{x}^i}\,\frac{\partial^2\bar{x}^i}{\partial x^b\,\partial x^c} = 0.$$

The functions $\gamma^i_k\,(x,\,dx)$ are the components of a geometrical object whose transformation law is given by (12·3).

Now suppose that $\gamma^i_k\,(x,\,dx)$ are linear homogeneous in dx, so that

$$(12\cdot4) \qquad \gamma^i_k\,(x,\,dx) = -\,\Gamma^i_{jk}\,dx^j,$$

where Γ^i_{jk} are functions of x alone. From (12·3) we have

$$(12\cdot5) \qquad \bar{\Gamma}^i_{jk} = \left(\Gamma^a_{bc}\,\frac{\partial x^b}{\partial\bar{x}^j}\,\frac{\partial x^c}{\partial\bar{x}^k} + \frac{\partial^2 x^a}{\partial\bar{x}^j\,\partial\bar{x}^k}\right)\frac{\partial\bar{x}^i}{\partial x^a}.$$

The geometric object of which the functions Γ^i_{jk} are the components is called an *affine connection* (see *Q. F.* Chap. III, §10). If Γ^i_{jk} are functions of class u' in the coordinate system x they will also be of class u' in $\bar{x}$, provided $u' \leq u - 2$. In this case u' is an invariant and may be called the class of the affine connection. When we refer to an affine connection it is to be assumed that $u' = u - 2$.

In case the second differentials with which we started satisfy the condition

$$(12\cdot6) \qquad \delta\,dx - d\,\delta x = 0,$$

the affine connection is symmetric, i.e.

$$\Gamma^i_{jk} = \Gamma^i_{kj}.$$

Conversely the symmetry of the affine connection implies (12·6).

The field of second differentials,

(12·7) $\delta dx^i = - \Gamma^i_{jk} dx^j \, \delta x^k$

defined by an affine connection is very special. There exists already a quite extensive theory of other geometric objects which arise by modifying the assumptions that lead to the affine connection. There are, for example, the fields of second differentials, $d\xi$, given by the equations (14·1) and (15·1) below. Another generalization is to require γ^i_k to be homogeneous of the first degree, but not necessarily linear, in dx. See the article by L. Berwald in the *Encyklopädie der Math. Wiss.* Vol. 3, part 3.

13. Parallel displacement.

Replacing dx^i by ξ^i in (12·7) and δ by d, we have

(13·1) $d\xi^i + \Gamma^i_{jk} \, \xi^j \, dx^k = 0.$

This together with any parameterized curve,

(13·2) $x^i = x^i(t), \qquad (b < t < c)$

where the functions $x^i(t)$ are of class 1, determines a set of ordinary differential equations,

(13·3) $$\frac{d\xi^i}{dt} = \gamma^i_j(t) \, \xi^j,$$

where

(13·4) $$\gamma^i_j(t) = - \Gamma^i_{jk}(x(t)) \, \frac{dx^k}{dt}.$$

By a standard existence theorem * (13·3) have a unique set of solutions

$$\xi^1(t), \, \ldots, \, \xi^n(t),$$

satisfying the initial conditions

$$\xi^i(t_0) = \xi^i_0,$$

with ξ^i_0 arbitrary and t_0 any number in the segment $b < t < c$.

From the linear character of the equations (13·3) it follows that these solutions are linear in the initial constants, that is to say the solutions are of the form

(13·5) $\xi^i(t) = a^i_j(t, t_0) \, \xi^j(t_0),$

---

* See G. A. Bliss, "Fundamental Existence Theorems" (*Colloquium lectures of the Amer. Math. Soc.* Vol. 3), p. 95, or L. Bieberbach, *Differentialgleichungen*, Berlin, 1926, p. 32. For a comprehensive treatment of parallel displacement making direct use of the linearity of equations (13·3), see L. Schlesinger, "Parallelverschiebung und Krümmungstensor," *Math. Annalen*, Vol. 99 (1928), p. 413.

and can be regarded as defining a linear homogeneous transformation

$$\xi(t_0) \rightarrow \xi(t)$$

from the tangent space at the point $x(t_0)$ to the tangent space at $x(t)$. By the existence theorem this holds in a certain maximum segment $b' < t < c'$.

The transformation $\xi(t_0) \rightarrow \xi(t)$ is non-singular for each pair of points in this segment, that is to say $a(t, t_0) \neq 0$, where $a = |a_j^i|$. For $(Q. F.$ Chap. I, § 7)

$$\frac{da}{dt} = a\alpha_i^s \frac{da_i^i}{dt},$$

where $a_s^i a_j^s = \delta_j^i$. But a_j^i are the solutions to (13·3) which reduce to δ_j^i for $t = t_0$. Therefore

$$\frac{da}{dt} = a\gamma_i^i(t),$$

and since $a(t_0, t_0) = 1$,

$$a(t, t_0) = e^{\int_{t_0}^{t} \gamma_i^i(s)\, ds} \neq 0.$$

From this it follows that $b' = b$ and $c' = c$. For assuming $c' < c$, we can start again at c' and define

$$\xi(c') \rightarrow \xi(t),$$

for some t such that $b' < t < c'$. The resultant of $\xi(t_0) \rightarrow \xi(t)$ and $\xi(t) \rightarrow \xi(c')$ exists, which contradicts the assumption $c' < c$. Therefore $c' = c$, and similarly $b' = b$.

The transformation (13·5) is independent of the coordinates because (13·1) is an invariant equation, and it follows from the form of (13·1) that (13·5) is independent of the parameterization of the curve (13·2). Therefore the transformation $\xi(t_0) \rightarrow \xi(t)$, of the tangent space at t_0 into the tangent space at t, is uniquely determined by the differential equations (13·1) and the curve (13·2). It is called the *parallel displacement* defined by Γ, of the first tangent space into the second along the curve (13·2) and corresponding vectors are said to be *parallel* with respect to this curve.

Since the set of solutions to (13·3) which satisfy given initial conditions is unique, it follows that:

If $\xi(t')$ and $\xi(t'')$ are both parallel with respect to (13·2) to a given vector $\xi(t)$, then $\xi(t'')$ is parallel to $\xi(t')$.

Letting $t'' = t$, we have:

If $\xi(t')$ is parallel to $\xi(t)$, then $\xi(t)$ is parallel to $\xi(t')$.

These conditions are expressed analytically by the functional relations

$$(13·6) \quad \begin{cases} (a) \quad a_j^i(t'', t) = a_s^i(t'', t')\, a_j^s(t', t), \\ (b) \quad a_s^i(t, t')\, a_j^s(t', t) = \delta_j^i, \end{cases}$$

which are necessary and sufficient conditions that the transformations (13·5) constitute a pseudo-group.

If the affine connection is flat (cf. Chap. III, § 10), vectors which are parallel with respect to any curve are equal and parallel, or equipollent, in the elementary sense. For in cartesian coordinates the components of Γ vanish, and (13·5) become

$$\xi(t) = \xi(t_0).$$

The analogy with the flat case is particularly close when the connection is symmetric. For in this case coordinate systems exist at which the components of the connection vanish at a given point, x ($Q.\ F.$ Chap. III, § 13). Let x_1 and x_2 be any points near x. Then the parallel displacement from x_1 to x_2 along the curve

$$x^i = x_1^i + t\,(x_2^i - x_1^i)$$

is approximately given in these coordinates by

(13·7) $$\xi^i(x_2) = \xi^i(x_1),$$

with an error of the second order as x_1 and x_2 tend to x.

This analogy has led to the term *infinitesimal parallelism*. When Γ is unsymmetric the analogy is lost, for the formulae corresponding to (13·7) are

(13·8) $$\xi^i(x_2) = \xi^j\{\delta_j^i - A_{jk}^{\ i}(x)\,(x_2^k - x_1^k)\},$$

where A is the alternating tensor, given by

$$A_{jk}^{\ i} = \tfrac{1}{2}\,(\Gamma_{jk}^i - \Gamma_{kj}^i).$$

In general the formulae (13·8) have nothing much to do with parallelism in the elementary sense. In a Euclidean space, for instance, the transformation (13·5), determined by the tensor A and the curve (13·2), might be the resultant of the translation in the underlying space (cf. § 6), which carries the point t_0 into t, followed by a rotation.

14. Affine displacement.

Any centred flat affine space, A^0, determines a unique flat affine space A, which has the same points and straight lines as A^0, the centre being regarded as equivalent to any other point. The affine space determined in this way by the tangent space of differentials at any point x, will be called the *tangent affine space* at x.

The transformations, defined by the equations for parallel displacement, between the tangent spaces of differentials at different points, are all isomorphisms between the tangent affine spaces. There are also families of displacements which are isomorphic transformations between the tangent affine spaces at different points, but which do not necessarily carry the null-vector of one space into the null-vector of the other. Such a family is defined by differential equations of the form

$$(14\cdot1) \qquad d\xi^i + \Gamma^i_{jk}\,\xi^j\,dx^k + B^i_k\,dx^k = 0,$$

where Γ is any affine connection, and B a tensor.

The equations $(14\cdot1)$ determine a unique transformation

$$(14\cdot2) \qquad \xi^i(t) = a^i_j(t, t_0)\,\xi^j(t_0) + a^i(t, t_0)$$

from the tangent space at a given point t_0, on the curve $(13\cdot2)$, to the tangent space at t. As in the case of parallel displacement, these transformations constitute a pseudo-group.

For let $h^i(t)$ be a particular set of solutions to the ordinary differential equations

$$(14\cdot3) \qquad \frac{d\xi^i}{dt} = \gamma^i_j(t)\,\xi^j + \gamma^i(t),$$

where $\gamma^i_j(t)$ are given by $(13\cdot4)$, and

$$(14\cdot4) \qquad \gamma^i(t) = -B^i_k(x(t))\frac{dx^k}{dt}.$$

By means of the substitution

$$(14\cdot5) \qquad \eta^i = \xi^i - h^i(t),$$

$(14\cdot3)$ are reduced to the equations $(13\cdot3)$, and we can apply the same arguments as in § 13.

Geometrically, this means that we apply the translation which carries $h(t)$ into the null-vector in the tangent space at each point of the curve $(13\cdot2)$.

The transformation $(14\cdot2)$ will be called the *affine displacement* defined by the composite geometric object consisting of Γ together with B, from the tangent space at t_0, along the curve $(13\cdot2)$, to the tangent space at t.

15. Generalizations.

An immediate generalization of the affine displacements is to consider the displacement of tangent spaces defined by

$$(15\cdot1) \qquad d\xi^i + \xi^j\,\Gamma^i_{jk}\,dx^k + B^i_k\,dx^k - \xi^i(\xi^j\,C_{jk}\,dx^k + D_k\,dx^k) = 0,$$

where Γ is an affine connection, and B, C, D are tensors of the kind indicated by the positions of the indices. This displacement is a projective transformation of the tangent spaces. To study it out we must introduce "points at infinity" in each tangent space and so define *tangent projective spaces*.

The theory of differentials is only one, arbitrarily chosen, method of attaching an associated space to each point of an underlying space (see Chap. VII, §§ 6–10). Such associated spaces arise naturally as a geometrical interpretation of many other geometric objects besides differentials. The totality of geometric objects with a given transformation law which are defined at a single point, x_0, is a space associated with x_0. This space satisfies the axioms given in Chap. II, § 9, G being the transformation law of the objects in question. Thus the totality of relative tensors which are covariant of order p, contravariant of order q, and have a given weight, is a space of n^{p+q} dimensions with a linear homogeneous transformation group. Likewise the totality of affine connections is an n^3-dimensional space with a group of linear transformations. These associated spaces are the final product of a process of evolution starting with the "systems of functions" introduced by Ricci (Q. F. Chap. II, § 16).

The idea of displacing associated spaces along curves (see Chap. VII, § 6), while it may also be regarded as a geometrical interpretation of processes already current in the theory of partial differential equations, had its real beginning with T. Levi-Civita's discovery of infinitesimal parallelism*, the full significance of which was brought out by H. Weyl†, who introduced the general symmetric affine connection. The idea of displacements between associated spaces was then taken up by E. Cartan and J. A. Schouten‡, who used it as a defining principle in generalized affine, conformal, and projective geometries. Displacements of a still more general type had previously been introduced by R. König§.

* "Nozione di Parallelismo in una varietà qualunque e conseguente specificazione geometrica della curvatura Riemanniana," *Rendiconti del Circolo Mat. di Palermo*, Vol. 42 (1917), pp. 173–204. Infinitesimal parallelism was also discovered independently by J. A. Schouten, "On the number of degrees of freedom of the geodetically moving system and the enclosing Euclidean space with the least possible number of dimensions," *Kon. Akad. van Weten. te Amsterdam*, Vol. XXI (1919), pp. 607–613.

† H. Weyl, "Reine Infinitesimalgeometrie," *Math. Zeit.* Vol. 2 (1918), pp. 384–411. See also H. Weyl, *Raum, Zeit, Materie*.

‡ E. Cartan, "Sur les variétés à connexion affine et la théorie de la relativité généralisée," *Annales de l'École Normale Supérieure*, Vol. 40 (1923), pp. 325–412; "Les espaces à connexion conforme," *Annales de la Soc. Polonaise de Math.* (1923), pp. 171–221, and "Sur les variétés à connexion projective," *Bull. de la Soc. Math. de France*, Vol. 42 (1924), p. 205; J. A. Schouten, "On the place of conformal and projective geometry in the theory of linear displacement," *Proc. Akad. van Weten. te Amsterdam*, Vol. 27 (1924), pp. 407–424, and "Erlanger Programm und Uebertragungslehre," *Rendiconti del Circolo Mat. di Palermo*, Vol. 50 (1926), pp. 142–169.

§ "Beiträge zu einer allgemeinen Mannigfaltigkeitslehre," *Jahresbericht der Deutschen Math. Vereinigung*, Vol. 28 (1920), pp. 213–228.

The working out of these conceptions has been linked up to a large extent with the development of the generalized projective geometry, of which the theory of projective displacements is one aspect, by the authors cited above, by T. Y. Thomas*, and by several others.

For an account of this subject and a bibliography see E. Bortolotti, "Connessioni proiettive," *Bollettino della Unione Matematica Italiana*, Vols. 9, 10 (1930–31).

* A projective theory of affinely connected manifolds, *Math. Zeit.* Vol. 25 (1926), p. 723.

CHAPTER VI

A SET OF AXIOMS FOR DIFFERENTIAL GEOMETRY

1. Purpose of the chapter.

We have now completed our account of simple manifolds of class u, having been mainly concerned with their local structure. The greater part of present-day differential geometry is the infinitesimal geometry of simple manifolds, and the previous chapters contain the elementary properties which are presupposed in most books on the subject.

But there are many spaces whose geometry can be studied by means of allowable coordinate systems, which do not satisfy the axioms of Chap. II, because there is no (1-1) continuous correspondence between the space and the arithmetic space. A projective space is a case in point. So is any closed sub-space, S_k, given by $(n-k)$ implicit equations (see Chap. IV, § 7) without singular points, for example a sphere or an anchor ring in the arithmetic 3-space.

The object of this chapter is to characterize a general class of spaces, of which S_k is a typical example. We do this axiomatically in terms of an undefined class of "allowable" coordinate systems, which are in many ways similar to the allowable coordinate systems for a simple manifold. But in general there is no allowable coordinate system in which the whole space is represented.

The axioms fall into three groups, A, B and C. The axioms A describe the local structure completely, and the axioms C impose certain general restrictions on the topology of the spaces*. The axioms B determine the class of allowable coordinate systems. Thus the structure is built up from the small to the large, in contrast to Chap. IV, §§ 3 and 7, where the local structure of a k-space is deduced from its representation as a whole.

The axioms describe a large class of spaces which are not all equivalent. In order to arrive at a particular geometry it is necessary to add further postulates of a more special nature.

* Any space satisfying the axioms A and C is a topological space with the same local structure as a simple manifold of class u. This, and other topological questions, are discussed in a paper called "A set of axioms for differential geometry," *Proc. Nat. Acad. of Sciences*, Vol. 17 (1931), p. 551, in which the independence of the axioms is also proved.

2. The first group of axioms.

The axioms will be stated in terms of "points" and "allowable coordinate systems." Points are completely undefined and allowable coordinate systems constitute an undefined class of (1-1) correspondences, $P \rightarrow x$, between sets of points and sets of arithmetic points in the arithmetic n-space.

The properties of the allowable coordinate systems will be described in terms of regular transformations of class u in the arithmetic space, and u is to be fixed, either as 0, 1, ..., ∞ or ω.

The image of an arithmetic n-cell in an allowable coordinate system will be called an n-cell of class u. As in the previous chapters we shall often omit the words "of class u" as applied to transformations and n-cells. The axioms of the first group are:

A_1. *The transformation of coordinates between two allowable coordinate systems which have the same domain is regular, provided one of them, at least, has a region for its arithmetic domain.*

A_2. *Any coordinate system obtained by a regular transformation of coordinates from an allowable coordinate system is allowable.*

A_3. *The correspondence in which each point of an n-cell corresponds to its image in an allowable coordinate system is an allowable coordinate system.*

3. The geometry of an n-cell.

We are now in a position to prove two theorems which justify our previous remark that the axioms A describe the local structure completely. The first theorem is:

Theorem 1. The image of an n-cell in any allowable coordinate system is an arithmetic n-cell.

Let X be the image of an n-cell, C, in an allowable coordinate system K. By the definition of an n-cell, C is the image in some allowable coordinate system, K_0, of an arithmetic n-cell, X_0. By A_3 the correspondences between C and X_0 in K_0, and between C and X in K, are allowable coordinate systems. By A_1 the transformation of coordinates from X_0 to X is regular, and therefore X is an arithmetic n-cell.

The second theorem is:

Theorem 2. If C is any n-cell and X any arithmetic n-cell, there exists an allowable coordinate system having C for its domain and X for its arithmetic domain.

By the definition of an n-cell, C is the image in some allowable coordinate system of an arithmetic n-cell, Y. By A_3 the correspondence between C

and Y is an allowable coordinate system. There is a regular transformation which carries Y into X, and the theorem follows from A_2.

The axioms A are sufficient to characterize an n-cell completely. For let an n-cell, C, be given and let us confine our attention to C. In other words let us add to the axioms A the further condition:

$\bar{A}$. *The space is an n-cell of class u.*

The arithmetic space is an arithmetic n-cell (Chap. III, § 6), and by theorem 2 at least one allowable coordinate system exists in which a space satisfying A and $\bar{A}$ is in (1-1) correspondence with the arithmetic space. With the set of all such allowable coordinate systems as preferred coordinate systems, an n-cell is obviously a simple manifold according to Chap. III, § 7. Conversely, it is obvious that a simple manifold with the allowable coordinate systems defined in Chap. III, § 9, satisfies the axioms A and $\bar{A}$.

4. The union of coordinate systems.

Let $[P]$ and $[Q]$ be the domains of (1-1) coordinate systems $P \to x$ and $P \to y$ respectively, $[x]$ and $[y]$ their arithmetic domains, and $[R]$

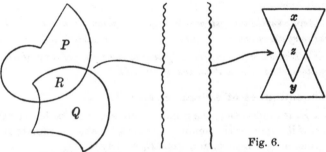

Fig. 6.

the intersection of $[P]$ and $[Q]$. It may happen that each R is represented by the same arithmetic point, z, in $P \to x$ as in $Q \to y$ and that $[z]$ is the complete intersection of $[x]$ and $[y]$, or else that $[R]$ and the intersection of $[x]$ and $[y]$ are both empty. In either case there is a [1-1] coordinate system in which each P corresponds to its image in $P \to x$, and each Q to its image in $Q \to y$. This coordinate system will be called the *union* of $P \to x$ and $Q \to y$.

Let $[K_a]$ be a set (finite or infinite) of (1-1) coordinate systems such that the union of each pair, K_a, K_β, exists. Let us denote the domain and arithmetic domain of K_a by U_a and X_a, respectively. Also let U denote the set of all points each of which is in at least one U_a, and let X denote the set of points each of which is in at least one X_a. Let K

be the correspondence in which each point of U corresponds to each arithmetic point by which it is represented in at least one K_α. If a point of U corresponded in this way to two different arithmetic points of X there would be two coordinate systems, K_α, K_β, in which this point corresponded to different arithmetic points, and the union of K_α and K_β would not exist. Hence each point of U corresponds in K to one, and only one, point of X. Similarly no two points of U correspond to the same arithmetic point, and therefore the correspondence K is (1-1). We call K the union of the set of coordinate systems $[K_\alpha]$. The theorem which we have proved is that any set of (1-1) coordinate systems, $[K_\beta]$, has a union if the union of each pair, K_α, K_β, exists.

5. The second group of axioms.

The purpose of the axioms B is to characterize the class of allowable coordinate systems in a space satisfying the axioms A. The section above describes the class of allowable coordinate systems whose domains are n-cells, and the axioms B say that the total class of allowable coordinate systems consists of the unions of these.

The axioms are:

B$_1$. *Any coordinate system which is the union of a set of allowable coordinate systems whose domains are n-cells, is allowable.*

B$_2$. *Each allowable coordinate system is the union of a set of allowable coordinate systems whose domains are n-cells.*

6. Consequences of axioms A and B.

From B_2 it follows that the union of any set of allowable coordinate systems, if it exists, is the union of a set of allowable coordinate systems whose domains are n-cells, and from B_1 we have:

Theorem 3. If the union of a set of allowable coordinate systems exists, it is an allowable coordinate system.

A set of points $[P]$ will be called a *region* if, and only if, each P is in an n-cell which is contained in $[P]$.

Theorem 4. The domain of an allowable coordinate system is a region.

For any allowable coordinate system, K, is the union of a set of allowable coordinate systems whose domains are n-cells, and each point in the domain of K is contained in one of these n-cells. Each of these n-cells is contained in the domain of K, which is therefore a region.

The image in an allowable coordinate system of any arithmetic region is obviously a region, and from theorem 1 it follows that a region which is contained in the domain of an allowable coordinate system is represented in this coordinate system by an arithmetic region. From theorem 4 we have, therefore:

Theorem 5. The arithmetic domain of an allowable coordinate system is a region.

Theorem 6. The correspondence between any region in the domain of an allowable coordinate system and its image in this coordinate system, is itself an allowable coordinate system.

Let X be the image of a region, U, in an allowable coordinate system, K. By the remark above, X is a region. Hence there is a set of boxes, $[X_a]$, each contained in X, and such that each point of X is contained in at least one X_a. If U_a is the image in K of X_a it follows from the definition of an n-cell that U_a is an n-cell, and from A_3 that the correspondence in K between U_a and X_a is an allowable coordinate system, K_a. It is obvious that each union, K_a, K_β, exists and therefore the union K' of $[K_a]$ exists, and is an allowable coordinate system by B_1. The arithmetic domain of K' is X and its domain is U.

Since the image in an allowable coordinate system of an arithmetic region is a region, we have:

Theorem 7. The correspondence between a region in the arithmetic domain of an allowable coordinate system, and its image in this coordinate system is itself an allowable coordinate system.

7. The third group of axioms.

C_1. *If two n-cells have a point in common they have in common an n-cell containing this point.*

C_2. *If P and Q are any two distinct points there is an n-cell C_P containing P, and an n-cell C_Q containing Q, such that C_Q has no point in common with C_P.*

C_3. *There exist at least two points.*

The relation between these axioms and those given by F. Hausdorff (*Grundzüge der Mengenlehre*, Leipzig, 1914, p. 213) for a topological manifold, is discussed in the note referred to in § 1 of this chapter.

8. Consequences of axioms A, B and C.

From C_2 and C_3 it follows that the space is a region. From C_1 we have:

Theorem 8. Two regions with a common point intersect in a region.

For let P be any point which is contained in each of two regions U and U'. By the definition of a region there are n-cells C and C' containing P, and contained in U and U', respectively. By C_1 there is an n-cell C'' containing P, and contained in the intersection of C and C', and therefore in the intersection of U and U'. Therefore the intersection of U and U' is a region.

From theorem 4 it follows that the domains of two allowable coordinate systems intersect in a region, if at all, and from theorem 6, theorem 5 and A_1 we have:

Theorem 9. *The transformation of coordinates between any two allowable coordinate systems, whose domains have a point in common, is regular.*

Theorem 10. *There is an allowable coordinate system in which two given points, P and Q, are both represented.*

By C_2 there are n-cells, C_P and C_Q, containing P and Q respectively, and having no point in common. Let C_x and C_y be any two arithmetic n-cells having no point in common. By theorem 2 there are allowable coordinate systems, K_x and K_y, in which C_P corresponds to C_x and C_Q to C_y, respectively. Neither C_P and C_Q nor C_x and C_y have points in common, therefore the union of K_x and K_y exists, and is an allowable coordinate system by B_1.

Theorem 11. *The totality of transformations between allowable coordinate systems is the pseudo-group of class u.*

By theorem 9 any transformation between two allowable coordinate systems belongs to the pseudo-group of class u.

Conversely any regular transformation $x \rightarrow y$, operating on an arithmetic region X, is the transformation between some pair of allowable coordinate systems. For there is at least one n-cell, and at least one allowable coordinate system in which a given n-cell is represented by the arithmetic space. From theorem 7 it follows that the region X is the arithmetic domain of some allowable coordinate system, and by A_2 that $x \rightarrow y$ is the transformation of coordinates between two allowable coordinate systems.

9. Manifolds of class u.

The general definitions and theorems which we have worked out in the last three chapters for simple manifolds, apply to any space satisfying the axioms A, B, and C. Nothing needs to be changed in the definition of k-cells, scalars and other geometric objects, nor in the discussion of tangent spaces, higher differentials, infinitesimal displacement, and so on. The same remark applies to the contents of $Q.F$.

A space which satisfies the axioms* A, B and C will be called *an*

---

* By defining a new class of allowable coordinate systems in terms of the old ones, if necessary, any space satisfying the axioms A and C alone can be made to satisfy the axioms A, B and C, and is therefore a regular manifold.

n-dimensional manifold of class u, or a *regular manifold*. A simple manifold of class u, as defined in Chap. III, is obviously a special case of a regular manifold. A manifold of class 0 satisfies our intuitive idea of continuity, and a manifold of class 1 our idea of "smoothness." The latter is expressed mathematically by the theory of first differentials.

Let U be any region in a regular manifold, M_n, and let the allowable coordinate systems for M_n, whose domains are contained in U, be taken as allowable coordinate systems for U. This set of allowable coordinate systems obviously satisfies the axioms A and B, and also the axioms C_1 and C_2. If U contains a single point it contains all the points in some n-cell, and therefore satisfies C_3. We have therefore:

Theorem 12. *Any region in a regular manifold is itself a regular manifold, provided it contains at least one point.*

Let $[K]$ be the totality of allowable coordinate systems for a manifold M_n, of class $u > 0$. Let $[K']$ be the set of coordinate systems, any one of which is obtained from a K by a transformation of class u', for any $u' < u$. With $[K']$ as allowable coordinate systems M_n obviously satisfies the axioms for a manifold of class u'. A manifold of class u is therefore a manifold of class u' with an additional element of structure, namely the sub-class of coordinate systems $[K]$. In fact the pseudo-group of class u is one of many which can be used to define particular classes of manifolds of class 0, just as the group G_u, defined in Chap. III, § 7, is one of many which can be used to define particular simple manifolds*.

Differential geometry may now be defined as the general theory of manifolds of class 1, as apart from the geometry of manifolds of class 0 which is a branch of Analysis Situs. The theory of tensors, tangent spaces, and smooth sub-spaces is present in all differential geometry. The general theory of manifolds of class 2 is a sub-class of differential geometries, which contain the theory of affine connections, curvature and osculating sub-spaces. Similarly there are differential geometries of class 3, class 4, ..., until we come to analytic manifolds, which have all the properties implied by power series expansions.

10. k-spaces in the large.

Theorem 13. *If P_0 is an ordinary point of $n - k$ scalars of class $u > 0$, $f^{k+1}(P)$, ..., $f^n(P)$, which are defined over a regular manifold, the set of ordinary points which satisfy the equations,*

$$(10\cdot1) \qquad f^\sigma(P) = f^\sigma(P_0), \qquad (\sigma = k+1, ..., n),$$

is a regular manifold of k dimensions.

* See Chap. VII, § 4, and also § 8 of the note referred to in § 1 above.

By the argument used in Chap. IV, § 3, it follows that each ordinary point on the locus (10·1) is contained in a k-cell consisting of points on the locus. Moreover, there is a class of parameterizations for each k-cell which are given in allowable coordinates by equations of the form

$$x^i = x^i (t^1, \ldots, t^k),$$

the functions $x(t)$ being of class u. If these parameterizations and their unions are taken as allowable coordinate systems the axioms A and B are obviously satisfied. From the axioms C for the n-space, it follows by the argument used in Chap. IV, § 7, that the axioms C are satisfied by the locus (10·1). Therefore the latter is a regular manifold of k dimensions.

11. Regular point transformations.

From theorem 10 it follows that the definition of point transformations given in Chap. III, § 15 is adequate for a discussion of local equivalence, where local equivalence on a regular manifold is defined as in Chap. IV, § 8. But for macroscopic equivalence we need a theory of point transformations between regions which are not necessarily represented in a single allowable coordinate system. This section contains the elements of such a theory.

With this as a basis it will be evident that §§ 16 and 17 of Chap. III may be applied to the theory of regular manifolds in general.

The general class of regular point transformations is defined in terms of a special sub-set which we define first. Let $P \to Q$ be a non-singular point transformation of an n-cell $[P]$ into some set of points $[Q]$, in the same or in a different regular manifold. The transformation $P \to Q$ will be described as *regular* if, and only if, there are allowable coordinate systems $P \to x$ and $Q \to x$, having $[P]$ and $[Q]$ respectively as their domains, having the same arithmetic domain, and such that $P \to Q$ is the resultant of $P \to x$ followed by $x \to Q$.

From this definition it follows that $[Q]$ is an n-cell. Therefore the inverse transformation $Q \to P$ is regular. It also follows from A_3 that the transformation $P \to Q$, operating on each n-cell contained in $[P]$ is regular. Moreover, a regular point transformation exists which carries one of two given n-cells into the other. For there is an allowable coordinate system in which a given n-cell corresponds to a given arithmetic n-cell.

A non-singular point transformation $P \to Q$, operating on a region $[P]$, will be called regular if, and only if, each P is contained in an n-cell

C_P, *which is contained in* $[P]$ *and is such that* $P \rightarrow Q$, *operating on* C_P, *is regular.*

From this definition it follows that $[Q]$ is a region and that $Q \rightarrow P$ is also regular. Let U be any region contained in $[P]$ and let P be any point in U. There exists an n-cell C_P, containing P, such that $P \rightarrow Q$, operating on C_P, is regular, and by theorem 8 there is an n-cell C_P' contained in the intersection of U and C_P. By a previous remark $P \rightarrow Q$, operating on C_P', is regular. Therefore $P \rightarrow Q$, operating on any region contained in $[P]$, is regular.

In case $[P]$ is an n-cell it is necessary to prove that the second of the definitions given above implies the first. This follows as a special case of the following theorem.

Theorem 14. *Let* $[P]$ *be the domain of any allowable coordinate system* $P \rightarrow x$, *and let* $P \rightarrow Q$ *be a transformation of* $[P]$ *into any set of points* $[Q]$. *The resultant,* $Q \rightarrow x$, *of* $Q \rightarrow P$ *followed by* $P \rightarrow x$, *is an allowable coordinate system if, and only if,* $P \rightarrow Q$ *is regular.*

First assume $Q \rightarrow x$ to be an allowable coordinate system, and let C_x be any n-cell in the arithmetic domain of $Q \rightarrow x$. The images of C_x in $P \rightarrow x$ and $Q \rightarrow x$, respectively, are n-cells, and it follows at once from A_3 that $P \rightarrow Q$ is regular.

Conversely if $P \rightarrow Q$ is regular there is an n-cell, C_P, containing a given P and contained in $[P]$, such that $P \rightarrow Q$, operating on C_P, is regular. That is to say there exist allowable coordinate systems, $P \rightarrow y$ and $Q \rightarrow y$, with the same arithmetic domain, which have as their respective domains C_P and the n-cell C_Q into which C_P is carried by $P \rightarrow Q$. Since $P \rightarrow x$ and $P \rightarrow y$ are allowable coordinate systems, the transformation, $x \rightarrow y$, between them is regular by theorem 9. Since $x \rightarrow y$ is regular, the coordinate system which carries each point in C_Q into its image in $Q \rightarrow x$ is allowable by A_2. Therefore $Q \rightarrow x$ is the union of allowable coordinate systems whose domains are n-cells, and is an allowable coordinate system by B_1.

As a corollary it follows that *a transformation,* $P \rightarrow Q$, *of a region* $[P]$ *into a region* $[Q]$ *is regular if, and only if,* $P \rightarrow Q$, *operating on each n-cell in* $[P]$, *is regular according to the first definition given in this section.*

We now show that the definition given in this section agrees with that given in Chap. III.

Theorem 15. *A point transformation,* $P \rightarrow Q$, *of a region* $[P]$ *into a set of points* $[Q]$, *where* $[P]$ *and* $[Q]$ *are both in the domain of a single* *allowable coordinate system, K, is regular if, and only if, it is repre-sented in K by a regular transformation in the arithmetic space.*

Let $[x]$ be the image of $[P]$ in K, $[y]$ the image of $[Q]$, and let $x \rightarrow y$ be the transformation by which $P \rightarrow Q$ is represented in K. First suppose $x \rightarrow y$ to be regular. Let $P \rightarrow x$ be the coordinate system in which each P corresponds to its image in K, and $Q \rightarrow y$ the coordinate system in which each Q corresponds to its image in K. By theorem 6, $P \rightarrow x$ and $Q \rightarrow y$ are allowable coordinate systems, and the transforma-tion $y \rightarrow x$ is a regular transformation of coordinates from $Q \rightarrow y$ to a coordinate system $Q \rightarrow x$, in which $[Q]$ is represented by $[x]$. The co-ordinate system $Q \rightarrow x$ is allowable by A_2 and is the resultant of $Q \rightarrow P$ followed by $P \rightarrow x$. By theorem 14 the transformation $P \rightarrow Q$ is regular.

Conversely let $P \rightarrow Q$ be regular. Then $Q \rightarrow x$ is an allowable co-ordinate system by theorem 14, and the transformation $x \rightarrow y$ is regular by A_1.

Theorem 16. A point transformation $P \rightarrow Q$, of a region $[P]$, into a set of points, $[Q]$, is regular if, and only if, for each P there is an n-cell C_P, containing P and contained in $[P]$, such that C_P and its image, C_Q, in $P \rightarrow Q$ are in the domain of a single allowable coordinate system, in which the transformation $P \rightarrow Q$ from C_P to C_Q is represented by a regular transformation in the arithmetic space.

This condition is sufficient by theorem 15 and the definition of a regular point transformation operating on a region. Therefore assume $P \rightarrow Q$ to be regular, and let P and Q be any pair of corresponding points. By theorem 10 there is an allowable coordinate system K, having P and Q in its domain. Let U be the domain of K, C_P' the intersection of $[P]$ with U, and let C_Q' be the image of C_P' in $P \rightarrow Q$. By theorem 8, C_P' is a region and therefore C_Q' is a region since $P \rightarrow Q$ is regular. Therefore the intersection of C_Q' and U is a region. Let C_Q be any n-cell containing Q and contained in this intersection, and let C_P be the image of C_Q in $Q \rightarrow P$. Then C_P is an n-cell, the regular transformation $P \rightarrow Q$ carries C_P into C_Q, and the theorem follows from theorem 15.

12. The pseudo-group of regular point transformations.

Let $P \rightarrow Q$ be a regular point transformation which carries an n-cell $[P]$ into an n-cell $[Q]$, and $Q \rightarrow R$ a regular transformation of $[Q]$ into an n-cell $[R]$. Then $P \rightarrow Q$ is the resultant of the transformations $P \rightarrow x$ and $x \rightarrow Q$, where $P \rightarrow x$ and $Q \rightarrow x$ are allowable coordinate systems with the same arithmetic domain $[x]$. Similarly $Q \rightarrow R$ is the resultant of the transformations $Q \to y$ and $y \to R$, where $Q \to y$ and $R \to y$ are allowable coordinate systems with the same arithmetic domain, $[y]$. Let $x \to y$ be the transformation of coordinates from $Q \to x$ to $Q \to y$. By A_1, $x \to y$ is regular, and by A_2 the coordinate system $P \to y$, obtained from $P \to x$ by the transformation $x \to y$, is allowable. Therefore the transformation $P \to R$, which is the resultant of the transformations $P \to y$ and $y \to R$, is regular. That is to say point transformations between n-cells have the transitive property.

Theorem 17. The totality of regular point transformations between regions in a regular manifold is a pseudo-group.

We have already remarked that the inverse of a regular point transformation is regular. Let $P \to Q$ be a regular point transformation which carries a region $[P]$ into a region $[Q]$, and $Q \to R$ a regular point transformation of $[Q]$ into a region $[R]$. By the corollary to theorem 14 $P \to Q$ carries any n-cell, C_P, in P into an n-cell, C_Q, by a regular transformation, and $Q \to R$ carries C_Q into an n-cell, C_R, by a regular transformation. From the transitivity of regular point transformations between n-cells it follows that $P \to R$ carries C_P into C_R by a regular transformation. Therefore $P \to R$, operating on the region $[P]$, is regular, and the theorem established.

CHAPTER VII

VARIOUS GEOMETRIES

1. Generalities.

In specializing from the general theory of the axioms A, B and C to some particular class of geometries, one may start by adding assumptions about the topology of the spaces, or one may specify some form of local structure. The one type of assumption may, to a certain extent, be made independently of the other. But many forms of local structure imply some restriction on the topological character of the space. For example, if n is even, a continuous vector-field without singular points cannot exist all over an n-sphere*.

One may also introduce some form of structure which implies both local and topological restrictions. A good example is to be found in the theory of continuous groups. An n-dimensional continuous group is a regular manifold with a continuous function, $F(P, Q)$, defined over all ordered pairs of points. The values of the function are points in the manifold, and it satisfies the conditions for a group given in Chap. II, § 2. For a brief account of this theory, and for references, see E. Cartan's Mémorial, *La théorie des groupes finis et continus et l'analysis situs*, Paris, 1930, No. XLII.

As another example let a set of n scalars $f^1(P)$, ..., $f^n(P)$, of class $u > 0$, having no singular point, be defined at each point of a regular manifold. These scalars determine a coordinate system, $P \to y$, given by

$$y^i = f^i(P),$$

in which a given point, P, has a single image, y. Several points may correspond to the same arithmetic point, but each P is contained in an n-cell which is carried by $P \to y$ into an arithmetic n-cell. Therefore the manifold is carried into an arithmetic region.

If we think of each point P as lying on the corresponding arithmetic point, the regular manifold may be one which overlaps itself in the manner indicated in fig. 7. Each point in the portion A of the diagram represents one point of the arithmetic space, but two points of the

---

* This theorem is due to L. E. J. Brouwer, "Ueber eineindeutige, stetige Transformationen von Flächen in sich," *Math. Annalen*, Vol. 69 (1910). A simplified proof has been given by J. W. Alexander, "On transformations with invariant points," *Trans. Am. Math. Soc.* Vol. 23 (1922), p. 94.

regular manifold. Such a manifold is said to lie smoothly on the arithmetic space.

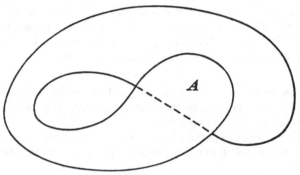

Fig. 7.

There can be no folding of the regular manifold on the arithmetic space. For a fold would imply an edge, i.e. a locus of singular points for the n scalars. Thus the regular manifold in the two-dimensional case could not be a sphere or an anchor ring or any other closed surface. Nor could it be obtained from an anchor ring by removing a finite number of isolated 2-cells.

2. The geometry of paths.

A space of paths is one in which there is, in addition to the properties of mere extension postulated in the axioms A, B and C, a system of curves, called paths, by means of which one may, so to speak, find his way about. There is a "best" path joining any two nearby points, P_0 and P_1, which may be given by a parametric equation of the form

$$P = F(P_0, P_1, t),$$

and we suppose $F(P_0, P_1, t)$ to be continuous in P_0, P_1 and t. The best path might well be taken as an undefined term in a system of axioms, but is generally defined by more specific conditions on the local structure. In a Riemannian space, for example, it is the shortest geodesic.

In building up a particular geometry of paths we may assume various topological properties, either in the large, e.g. that any two points are joined by at least one path, or in the small, e.g. that a 2-cell, given by

$$P = F(F(O, A, u), F(O, B, v), w),$$

with suitable restrictions on u, v and w, is generated by the best paths joining points on two given paths with a common point.

Other assumptions may refer to classes of parameterizations for the individual paths. Thus an affine space of paths is one in which every path has a local structure defined by a family of affine parameterizations for each of its 1-cells. The affine parameterizations are those, and only those, obtained from a given one by linear equations

$$s^* = as + b.$$

Thus each path is a locally Euclidean 1-space, for which the affine parameterizations are locally cartesian coordinate systems (cf. Chap. II, § 13; Chap. III, § 12, and § 3 below).

The affine spaces of paths which have been most widely studied are those in which the paths are given in allowable coordinates by the differential equations

$$(2\cdot1) \qquad \frac{d^2 x^i}{ds^2} + \Gamma^i_{jk} \frac{dx^j}{ds} \frac{dx^k}{ds} = 0,$$

where Γ^i_{jk} are the components of a symmetric affine connection. The existence of best paths follows from the fact that such a space is locally convex. That is to say, if P is a given point in a given region U, there is an n-cell C_P, containing P and contained in U, such that two given points in C_P are joined by one, and only one, arc of a path which does not leave† C_P.

The equations

$$(2\cdot2) \qquad \xi^i_{,k} = 0,$$

where ξ is a contravariant vector, and the comma denotes covariant differentiation with respect to Γ, may be interpreted in terms of an infinitesimal parallelism as in Chap. V, § 13. The tangents at different points of a path are parallel with respect to the path. Other parallel displacements for which the paths are the self-parallel curves arise as the geometrical interpretation of differential equations

$$(2\cdot3) \qquad \xi^i_{,k} + A^i_{jk} \xi^j = 0,$$

where A is a tensor which is alternating in its covariant indices. Similarly affine displacements may be defined by means of the equations

$$(2\cdot4) \qquad \xi^i_{,k} + A^i_{jk} \xi^j + B^i_k = 0,$$

where B^i_k is any mixed tensor.

The theory of a particular family of affine displacements belongs to the geometry of the composite geometric object consisting of the sym-

† J. H. C. Whitehead, "Convex regions in the geometry of paths," *Quarterly Journal of Math.* (Oxford series), Vol. 3 (1932), p. 33.

metric affine connection together with the tensors A and B. Its relation to the geometry of T alone is analogous to that which Euclidean geometry bears to affine geometry. Moreover the additional elements of structure are defined by tensors in both cases.

Symmetric affine connections have been studied in some detail when $u = \omega$. The elementary theory of normal coordinates and normal tensors is to be found in Q. F. Chap. VI. Many results are independent of the restriction $u = \omega$, and a brief discussion of normal coordinates for $u \leqq \omega$ is to be found in the note referred to in Chap. VI, § 1.

These geometries have been generalized by allowing the components Γ^j_{jk} in (2·1) to depend on the direction dx^1, ..., dx^n as well as on the point x. They are therefore homogeneous of degree zero in the variables dx^1, ..., dx^n. For an account of this theory the reader is referred to J. Douglas, "The general geometry of paths," *Annals of Math.* Vol. 29 (1928), p. 143.

As yet there does not exist a systematic geometry of paths of class $u = 0$, 1 or 2, which must get along without the formalism available when $u > 2$. But the beginnings of this geometry are to be found in the "Textile mathematics" of W. Blaschke and his school, a general account of which may be found in the article by W. Blaschke, "Neue Strömungen der Differentialgeometrie," *Jahresbericht der Deutschen Math. Ver.* Vol. 40 (1931), p. 1.

3. Locally flat affine spaces.

The definitions and results of Chap. III, § 12 can be extended in a few words to regular manifolds in general. A regular manifold over which a locally flat affine connection is defined will be called a *locally flat affine space*. Thus a locally flat affine space is an affine space of paths according to the last section.

The locally cartesian coordinates are defined as in Chap. III, § 12, and each point is represented in at least one locally cartesian coordinate system. Moreover, the locally flat cartesian coordinate systems obviously satisfy the axioms A and B if the word "linear" is substituted for "regular" in A_1 and A_2.

Conversely, let us substitute in the axioms A and B the word "linear" for "regular" and "locally cartesian" for "allowable." We thus obtain new sets of axioms, A and B, which, together with C as they stand, form a consistent set. For they are satisfied by a flat affine space. From the formal argument used in Chap. III, § 10, it follows that a locally flat affine connection is defined all over any space satisfying these axioms.

Therefore *the class of locally flat affine spaces is the class of spaces which satisfy the axioms obtained by substituting for the pseudo-group of class u in A, B and C the pseudo-group of linear transformations operating on arbitrary regions.*

Locally Euclidean spaces may be defined in the same way, as well as spaces of constant positive or negative curvature. For a very clear account of the history and significance of these spaces, with references to the work of W. K. Clifford, W. Killing, and himself, see F. Klein, "Zur Nicht-Euklidischen Geometrie," *Math. Annalen*, Vol. 37 (1890), or *Gesammelte Mathematische Abhandlungen*, Berlin, 1921, Vol. 1, p. 353. For modern work on the same series of problems, see H. Hopf, "Zum Clifford-Kleinschen Raumproblem," *Math. Annalen*, Vol. 95 (1925), p. 313, and E. Cartan, *Leçons sur la géométrie des espaces de Riemann*, Paris, 1928 (Chap. III).

4. Other pseudo-groups.

The characterization of locally flat affine spaces by means of a set of axioms similar to A, B and C, but using a sub-pseudo-group of the pseudo-group of class u, is an example of a general method by which particular differential geometries can be defined. It is analogous to the passage from the geometry of a group as defined in Chap. II, § 9, to the geometry of a sub-group.

Thus there is a class of spaces (locally flat projective spaces) defined by the pseudo-group of linear fractional transformations between regions. Similarly, the pseudo-groups of conformal, of contact, of volume preserving, and of many other types of transformations between regions, give rise to classes of spaces which deserve study. A few of these are mentioned in the paper referred to in Chap. VI, §1.

5. Oriented manifolds.

Any manifold determined by the pseudo-group of direct transformations of class $u > 0$ (Chap. III, §8), will be called an *oriented manifold* of class u. A regular manifold is called *orientable* if there is a sub-set of its allowable coordinate systems, associated with which it satisfies the axioms for an oriented manifold. In § 10 below it is shown that if an orientable manifold is connected, i.e. does not consist of two regular manifolds having no common point, there are two, and only two, such sub-sets of its allowable coordinate systems. The coordinate systems in one of these sub-sets may be arbitrarily described as *positively oriented*. This is called *orienting* the manifold.

A simple manifold is orientable as follows from Chap. III, §8.

6. Displacements of associated spaces.

A large class of geometries are concerned with the theory of displacements of associated spaces to which reference has been made at the end of Chap. v. These displacements are generalizations of the affine displacements described in Chap. v. The general process may be described in abstract terms as follows:

With each point P of an underlying regular manifold there is associated a space $S(P)$, and all these spaces are isomorphic. A family of displacements is a set of transformations, $S(P) \to S(Q)$, such that:

(1) Any displacement $S(P) \to S(Q)$ is an isomorphic transformation of $S(P)$ into $S(Q)$.

(2) If P and Q are any two points of the underlying manifold there exists at least one displacement $S(P) \to S(Q)$.

(3) The resultant of a displacement $S(P) \to S(Q)$ followed by a displacement $S(Q) \to S(R)$ is a displacement $S(P) \to S(R)$.

(4) The inverse of any displacement $S(P) \to S(Q)$ is a displacement $S(Q) \to S(P)$.

Thus a family of displacements is a pseudo-group of isomorphisms between associated spaces. The standard method of determining a family of displacements is by means of a geometric object, such as an affine connection, which, together with a curve joining two points P and Q, determines a transformation $S(P) \to S(Q)$. This was worked out in Chap. v for parallel and affine displacements in a simple manifold. We propose now to show how this argument can be extended* to an arbitrary connected manifold, formulating the discussion so that it applies to any geometric object which has certain abstract local properties in common with an affine connection.

Let a space $S(P)$ be associated with each point P of the underlying manifold. Let Δ be a geometric object which determines a pseudo-group of displacements along any 1-cell of class† u', and therefore along any arc (by which we mean a closed interval in a 1-cell, cf. Chap. ii, §12).

We shall show that Δ determines a pseudo-group of displacements along any curve consisting of a finite number of arcs, $\sigma_1, \dots, \sigma_N$, the last point of σ_α being the first point of $\sigma_{\alpha+1}$. Such a curve may be

---

* The essential generalization is to curves which are not necessarily represented in any single allowable coordinate system.

† The class u' is 1 for affine displacements, 2 for displacements defined by differential equations involving second derivatives of the functions defining the 1-cell, and so on.

represented as the sum of the arcs, thus

$$\gamma = \sigma_1 + \ldots + \sigma_N.$$

It may have any number of multiple points.

The theorem follows by an induction on the number N. Assuming it true for the sum, γ_{N-1}, of any $N-1$ arcs, we have to prove it for any curve

$$\gamma = \gamma_{N-1} + \sigma$$

which can be represented as the sum of N arcs. Let A be the last point of γ_{N-1} and the first point of the arc σ. Then the displacement $S(P) \to S(P')$, where P and P' are any points on γ, may be defined as the resultant of $S(P) \to S(A)$, followed by $S(A) \to S(P')$. The set of all such displacements is obviously a pseudo-group. It follows from the hypothesis of the induction and the original assumptions about Δ, that $S(P) \to S(P')$ is an isomorphic transformation of $S(P)$ into $S(P')$. Therefore the theorem follows from the fact that it is true for $N = 1$.

Any two points in a connected manifold are joined by at least one such curve, and therefore Δ determines a pseudo-group of displacements over the manifold if the latter is connected.

7. The holonomic group.

If P is any point in a manifold over which a pseudo-group of displacements is defined, the set of displacements which carry $S(P)$ into itself is obviously a group contained in the pseudo-group. It is a subgroup of the group of automorphisms of $S(P)$, and is called the *holonomic* group* at P. The holonomic groups at any two points P and Q are simply isomorphic. For there is a displacement $S(P) \to S(Q)$, and the two groups are obviously conjugate under any such displacement.

If, in particular, the holonomic group at any one point reduces to the identity, it reduces to the identity at every point, and the displacement is said to be *holonomic*. This is the case for example with parallel displacements in a flat affine geometry (cf. Chap. V, §13). The displacement is necessarily holonomic if the group of automorphisms of $S(P)$ is the identity.

When the displacement is holonomic, there is a unique transformation, $S(P) \to S(Q)$. For if there were two distinct displacements of $S(P)$ into $S(Q)$, the resultant of either one followed by the inverse of the other would be a transformation, other than the identity, of the holonomic group at P.

---

* See E. Cartan, "Les groupes d'holonomie des espaces généralisés," *Acta Math.* Vol. 48 (1926), p. 1.

If the displacements $S(P) \rightarrow S(Q)$ are defined along curves having P and Q as their end-points, the holonomic group at P_0 consists of the displacements round closed curves,

$$P = P(t), \qquad (t_0 \leqq t \leqq t_1)$$

where $\qquad\qquad P(t_0) = P(t_1) = P_0.$

8. Locally holonomic displacements.

Let U be any region in a manifold over which there is a family of displacements along curves. This family contains a sub-pseudo-group consisting of all the displacements along curves which are contained in U. If this sub-pseudo-group of displacements is holonomic the original family may be described as holonomic in U. The family will be described as *locally holonomic* if, and only if, each point is contained in an n-cell in which the displacement is holonomic. Thus the parallelism defined by a locally flat affine connection is locally holonomic*.

9. The holonomic group of an affine displacement.

Let $G(P)$ be the holonomic group at a point P, of the family of affine displacements determined by the equations (2·4). A transformation of $G(P)$ is obviously given by equations of the form

$$(9\cdot1) \qquad\qquad \bar{\xi}^i = a^i_j(\gamma)\, \xi^j + a^i(\gamma),$$

where $a^i_j(\gamma)$ and $a^i(\gamma)$ are functions of the closed curves which begin and end at P. Since $G(P)$ is a group it follows that

$$(9\cdot2) \qquad a^i_j(\gamma_1 + \gamma_2) = a^i_s(\gamma_2)\, a^s_j(\gamma_1),$$

$$a^i(\gamma_1 + \gamma_2) = a^i_s(\gamma_2)\, a^s(\gamma_1) + a^i(\gamma_2),$$

where $a(\gamma_1 + \gamma_2)$ are the coefficients which define the displacement round the curve $\gamma_1 + \gamma_2$.

It can be proved that *the coefficients* $a(\gamma)$ *are continuous functions of the curve* γ. Also that *the set of transformations given by* (9·1), *where* γ *is a curve which can be continuously deformed into* P, *is an invariant sub-group of* $G(P)$.

If γ is given in a coordinate system x by the quadrangle whose vertices are x, $x + dx$, $x + dx + \delta x$, and $x + \delta x$ respectively, the transformation (9·1) may be written

$$(9\cdot3) \qquad\qquad \bar{\xi}^i = \xi^i - \xi^j R^i_{jkl}\, dx^k\, \delta x^l - R^i_{kl}\, dx^k\, \delta x^l + \epsilon^i,$$

---

* A Möbius band with a locally flat affine structure is an example of a locally flat affine space on which the parallelism is not holonomic. (See p. 319 of the paper by Hopf referred to in § 3.)

where ϵ^i is a small quantity of the third order in dx and δx, and

$$(9\cdot4) \qquad \begin{cases} R^i_{jkl} = B^i_{jkl} + A^i_{jkl}, \\ R^i_{kl} = B^i_{k,l} - B^i_{l,k} + B^s_k A^i_{sl} - B^s_l A^i_{sk}. \end{cases}$$

In these formulae B^i_{jkl} are given by $(12\cdot2)$ Chap. III, A^i_{jkl} by the same formulae with A^i_{jk} substituted for Γ^i_{jk}, and the comma denotes covariant differentiation with respect to Γ. This follows by arguments similar to those used by Eisenhart, *Non-Riemannian geometry*, Chap. I, §10. The calculations can be simplified by using normal coordinates for Γ. The tensors R^i_{jkl} and R^i_{kl} are called the tensors of *curvature* and *torsion*, respectively, and play a central rôle in the researches covered by references to Cartan and others in Chap. V, §15.

10. Displacement of orientation.

The tangent space of differentials $T(P)$, at any point P, determines two oriented tangent spaces, $T_1(P)$ and $T_2(P)$. There is thus an associated space $S(P)$, which contains just two points, namely $T_1(P)$ and $T_2(P)$. An orientation of the tangent space at any point P determines a unique orientation of any n-cell, C_P, containing P, and this orientation of C_P determines an orientation of the tangent space at each point in C_P (Chap. V, §5). Therefore an orientation of $T(P)$ determines an orientation of the tangent space at each point in C_P, and in particular at each point P' of any 1-cell, σ, containing P and contained in C_P. If two n-cells determine the same oriented tangent space at a common point, P, they obviously determine the same orientation of any n-cell containing P and contained in their intersection. It follows by a simple argument that the orientations of the tangent spaces at points of the 1-cell, σ, are determined independently of the particular n-cell, C_P, by a given orientation of $T(P)$. That is to say a pseudo-group of displacements is defined between the two-point spaces, $S(P')$, associated with points of σ, and hence a pseudo-group of displacements is defined over the whole manifold, provided the manifold is connected. The displacements are between the associated spaces $S(P)$, and are along curves of the sort considered in §6.

It is obvious from the definition that this family of displacements is locally holonomic. *It is holonomic if, and only if, the underlying manifold is orientable.*

Proof. If the manifold is orientable a unique orientation can be assigned to every n-cell in such a way that two n-cells with a common point, P, determine the same orientation of $T(P)$. Therefore a unique orientation is assigned to every tangent space by the method used in defining the displacement, and the latter is holonomic.

Conversely, if the displacement is holonomic, a unique orientation, $T_1(O)$, of the tangent space at a given point O, determines a unique orientation, $T_1(P)$, of the tangent space at each point P. Let P and P' be two points in any n-cell, C. The oriented tangent space $T_1(P)$ determines an orientation of C, which is the same as that determined by $T_1(P')$, since the displacement is holonomic. Therefore a unique orientation is assigned to every n-cell, which agrees with the orientation, $T_1(P)$, of the tangent space at any point in the n-cell.

The underlying manifold, associated with the positively oriented coordinate systems for these oriented n-cells, obviously satisfies the axioms A_1, A_2 and C for an oriented manifold. The axiom A_3 is satisfied because any two n-cells with a common point, P, determine the same orientation of $T(P)$, namely $T_1(P)$, and so they determine the same orientation of any n-cell containing P and contained in their intersection. Therefore the underlying manifold is orientable.

It follows that a connected oriented manifold carries two, and only two, oriented manifolds, as stated in § 5.

The two-point associated spaces may be thought of as a notational device. Essentially the same argument as above may be made by observing that the way in which the orientation of a tangent space $T(P)$, at any point of an n-cell C, determines an orientation of the tangent space at any other point of C, is by letting each differential at P correspond to the differential at Q which has the same components in some allowable coordinate system whose domain is the cell C. This amounts to displacing $T(P)$ to $T(Q)$ by the locally flat affine connection whose components are zero in this coordinate system. We are thus considering the pseudo-group of all parallel displacements which can be affected between tangent spaces by locally flat affine connections. The holonomic group of this family of displacement is either the group of direct linear homogeneous transformations of a tangent space into itself, or the whole centred affine group. In the first case the underlying manifold is orientable, in the second case not.

A locally holonomic displacement on a manifold in which any closed curve is deformable into a point is obviously holonomic. It follows that such a manifold is orientable.

11. Covering manifolds.

The associated spaces, $S(P)$, need have no structure beyond their cardinal number. They may then be regarded as the sets of values of some function*$\xi(P)$. Let Δ be a locally holonomic family of displacements between such a set of associated spaces, which is defined over a regular manifold M_n. Let U be any region in which Δ is holonomic, P_0 a fixed and P a variable point in U. Let $\xi'(P_0)$ be a given value of $\xi(P_0)$ and $\xi'(P)$ the value of $\xi(P)$ into which $\xi'(P_0)$ is displaced by Δ. The single-valued function $\xi'(P)$ is called a branch of $\xi(P)$. The number of these branches is the cardinal number, α, of each associated space $S(P)$.

Let a new manifold, $\overline{M}_n$, be defined as follows. A point, $[P, \xi(P)]$, in $\overline{M}_n$ is to be a point, P, in M_n, associated with a point, $\xi(P)$, in $S(P)$. The coordinate system

$$[P, \xi'(P)] \to x$$

is to be an allowable coordinate system for $\overline{M}_n$, where $P \to x$ is any allowable coordinate system for M_n over whose domain Δ is holonomic, and $\xi'(P)$ is one of the single-valued branches described above. With this family of allowable coordinate systems $\overline{M}_n$ obviously satisfies the axioms for a regular manifold. To each point of $\overline{M}_n$ corresponds just one point of M_n, and to each point of M_n correspond α points of $\overline{M}_n$. The manifold $\overline{M}_n$ is said to cover M_n α times.

The above transformation of M_n into $\overline{M}_n$ carries $\xi(P)$ into a single-valued function defined over $\overline{M}_n$, namely the function whose value at $[P, \xi'(P)]$ is $\xi'(P)$.

When $S(P)$ consists of the two oriented tangent spaces $T_1(P)$ and $T_2(P)$ the covering manifold $\overline{M}_n$ is orientable. For, by definition, a closed curve on $\overline{M}_n$ is one which begins and ends with a given point, P, associated with the same orientation of $T(P)$, and it follows that the displacement of orientation in $\overline{M}_n$ is holonomic. $\overline{M}_n$ is a connected manifold if M_n is connected and not orientable, and is a pair of connected manifolds if M_n is connected and orientable.

---

* The values of this function may or may not be numbers.

INDEX OF DEFINITIONS